Cheorl-Ho Kim

Glycosphingolipids Signaling

Springer

Cheorl-Ho Kim
Molecular and Cellular Glycobiology
Unit, Department of Biological Sciences
SungKyunKwan University
Suwon, Korea (Republic of)

ISBN 978-981-15-5806-1 ISBN 978-981-15-5807-8 (eBook)
https://doi.org/10.1007/978-981-15-5807-8

This Springer imprint is published by the registered company Springer Nature Singapore Pte Ltd.
The registered company address is: 152 Beach Road, #21-01/04 Gateway East, Singapore 189721, Singapore

Keywords: *Sialobiology; Ganglioside; Neuraminic acid; Glycan receptor; Glycan ligand; Cellular function; Infection; Transformation*

Preface

Glycoconjugates are abundant on the surface of animal cells. Membrane glycoconjugates can mediate intercellular adhesion, interactions, recognition, and signaling. Diverse types of glycolipids are involved in cellular processes, primarily depending on structural and distribution properties because glycolipids form functional membrane bilayers in cellular organelles. Glycolipids are a group of glycosides linked to amino alcohols, alcohols, sterols, and fatty acids. Glycolipids are therefore classified to a group of glycoconjugates and mainly localized on the cellular surfaces of prokaryotic bacteria and eukaryotic kingdom. In plants and prokaryotic bacteria, which are distinctly different from animals, glycoglycerolipids designate their major type of glycolipid species. Current knowledge of the lipids is mainly focused on their roles in bioactive second messengers which have subjected for phosphorylation and intracellular signal transduction during the last three decades. Cellular plasma membranes (PMs) are phosphor- and glycolipid bilayers easy to compartment due to their mosaic pattern. Then, the polar groups are located on the asymmetrical surface. In cells, membrane lipids are found as forms of cholesterol, phosphatidylserine, sphingomyelin, phosphatidylcholine, and phosphatidylethanolamine in the structural quality and quantity. However, lipids present in relatively low quantities are ceramides, diacylglycerol, phosphatidic acid, or phosphoinositides, but they are used as signaling molecules when they interact with protein targets. For functional significance of glycolipids, the word glycol, attached to lipids, denotes the glycans or carbohydrates.

Glycosphingolipids (GSLs) are mainly anchored into cellular membrane and therefore components of complex carbohydrate moieties linked to hydrophobic ceramides (Cers) as embedded lipophilic backbone. As a third life chain, carbohydrates encode information in glycosidic linkages [1], and the carbohydrate information is phenotypically expressed upon interaction with each specific counterpart such as lectins and receptors. Many GSLs assemble into discrete microdomains known as lipid rafts and act as membrane communicators between extracellular and intracellular world. In fact, they are binding sites for invasive pathogens and also receptors for bacterial toxins. Representative example is the lectin that binds to the carbohydrate ligands [2].

GSLs are pathologically involved in the various human diseases such as glycosphingolipidoses, peripheral neuronal immunopathies, GSLs-reactive antibodies-causing diseases, toxin-mediated diarrhea. Some important genetic diseases, sphingolipidoses, are associated with defects in the glycolipid catabolism pathway. Therefore, GSL structures, nomenclature, and metabolism are important for understanding of their functions. GSLs also function as human alloantigens, which blood group AB(H)O, I/I, Lewis, and PP1Pk are included, and as xenoantigens such as Forssman antigen and galactose (Gal) antigen. Human anti-ganglioside antibodies therefore induce autoimmune diseases. To date, cellular GSLs are associated with many biological functions such as cellular oncotransformation, phenotype change, neuronal or embryonic development, cell division regulation, cell–cell interaction, attachment, adhesion and motility as well as intracellular signaling through carbohydrate-to-carbohydrate interaction (CCI) or protein-to-carbohydrate interaction (PCI). Understanding of the precise roles of neuraminyl or sialyl glycans is needed. The sialoglycoproteins or sialoglycolipids present in the multicellular system will certainly open new strategic challenges against emerging infection agents or intractable diseases which are emerged with co-evolutionary development.

Suwon, Korea Cheorl-Ho Kim

References

1. Manning JC, Romero A, Habermann F, García Caballero G, Kaltner H, Gabius HJ (2017) Lectins: a primer for histochemists and cell biologists. Histochem Cell Biol 147(2):199–222
2. Rüdiger H, Gabius HJ (2009) The biochemical basis and coding capacity of the sugar code. In: Gabius HJ (ed) Fundamentals of glycosciences. Wiley, Weinheim, pp 3–14

Contents

Abbreviations

GSL	Glycosphingolipid
CDw17	LacCer
ES	embryonic stem
UGCG	UDP-glucose:ceramide glucosyltransferase
CCI	carbohydrate-carbohydrate interaction
GalT-1	galactosyltransferase I
GFR	growth factor receptor
GM2 synthase	b-1,4 N-acetylgalactosaminyltransferase 1 (GalNAc transferase)
GM3 synthase	ST3GalV/SAT-1
PM	plasma membranes
RTK	receptor tyrosine kinase, ST3GalV/SAT-1
SSEA	stage-specific embryonic antigen
SSEA-4	sialyl-globopentaosyl-Cer
SSEA-5	Fuc-Lac-traosyl-Cer
MAb	monoclonal antibody
LabA	LacdiNAc-binding adhesin
SabA	SA-binding adhesin
HPNAP	*H. pylori* recognizing neutrophil-activating protein
Pla	plasminogen activator
SEB	Staphylococcal enterotoxin B
BoTx	botulinum toxin
BoNT	botulinum neurotoxin
GM-CSF	granulocyte-macrophage colony-stimulating factor
GMR	GM-CSF receptor
HBGA	histo-blood group antigen
SL	sialyllactose
GBS	Guillain-Barré syndrome
CNS	central neuronal system

NSC	neural stem cells
PD	Parkinson's disease
HD	Huntington's disease
Trk	tropomyosin-related kinase
GAG	glycosaminoglycan
HS	heparan sulfate
KS	keratan sulfate
BMP	bone morphogenic protein
RIG-I	retinoic acid (RA)-inducible gene I
MDA	melanoma-differentiation-associated protein
ARF	ADP ribosylation factor
LFA	leukocyte function-associated antigen
TeNT	tetanus neurotoxin
BoNT	botulinum neurotoxins
SNARE	N-ethylmaleimide-sensitive factor (NSF) attachment protein receptor
BabA	blood group-binding adhesin
GIT	gastrointestinal tract
LabA	LacdiNAc-binding adhesin
STAT3	signal transducer activator of transcription
ILF	Leukemia inhibitory factor
ESC	embryonic SC
iPSC	induced pluripotent stem cells
TRA	tumor recognition antigen
ECC	embryonal carcinoma cell
ST	sialyltransferase
TAMs	tumor-associated microenvironments
TSP	tetraspanin
Trk	tyrosine kinase receptor
FGFR	fibroblast growth factor receptor
EGFR	epidermal growth factor receptor
bFGFR	basic FGFR
PDGFR	platelet-derived growth factor receptor
NGFR	nerve growth factor receptor
IGFR	insulin-like growth factor receptor
IR	insulin receptor
PSI	protein-sphingolipid interaction
VEGFR	vascular endothelial growth factor receptor
GDNF	glia cell-derived neurotrophic factor
B2R	bradykinin 2 receptor
FasR	CD95
GPCR	G protein-coupled receptor
SR	serotonin receptor

AR	α1-adrenergic receptor
Aβ	amyloid-β
ER	estrogen receptor
HGFR	hepatocyte growth factor receptor
SP	sphingosine
TGFβ	transforming growth factor-β
AMPA	α-amino-3-hydroxy-5-methylisozazole-4-propionic acid
AMPAR	ionotropic glutamate receptor
LTP	ATP-elicited long-term potentiation
HFS	high-frequency electric stimulation
BDGF	brain-derived neurotrophic factor
NMDAR	N-methyl-D-aspartate receptor
ERK	extracellular signal-regulated kinase
CREB	cAMP responsive element binding protein
CRE	cAMP response element
PLC	phospholipase C
mGluR	metabotropic glutamate receptor
IP3	inositol triphosphate
DAG	diacylglycerol
PtdIns4,5P2	phosphatidylinositol 4,5-bisphosphate
VCAM	vascular cell adhesion molecule-1
ICAM	intracellular CAM
NCAM	neural CAM
EMT	epithelial-to-mesenchymal transition
GBM	GSL-binding motif
SM	sphingomyelin
STx	Shiga toxin
5-HT	5-hydroxytryptamine CTx, cholerae toxin
LT	heat labile toxin
CTA	CTx enzymatic A-subunit
CTB	CTx B-subunits
TeNT	tetanus neurotoxin
PTx	pertussis toxin
Mac-1	CD11b/CD18
TLR	Toll-like receptor
TORAP	Toll-IL-1R domain-containing adaptor protein
BabA	blood group-binding adhesin
GIT	gastrointestinal tract
ETCE	**enterotoxigenic *E. coli***
UTEC	**uropathogenic *E. coli***
STCE	ST-producing *E. coli*

VT	Verotoxin
Asialo-GM1 (Gg4Cer)	Galβ1,3GalNAcβ1, 4Galβ1,4Glc-Cer
Asialo-GM2 (Gg3Cer)	GalNAcβ1,4Galβ1,4Glc-Cer
Lacto-neotetraosyl-Cer	nLc4-Cer
Sialyl-lacto-tetraose	s-Lc4
Globo H	SSEA-3, globopentaosyl-Cer
Gg3	gangliotriaosyl-Cer
Gb3	CD77, pk blood group antigen
globotriaosyl-Cer	Galα1,4Galβ1, 4Glcβ1,1Cer
Gb4	globotetraosyl-Cer: GalNAcβ1,3Galα1,2Galβ1, 4Glcβ1,1-Cer
Forssman antigen globoseries	GalNAcα1,3GalNAcβ1, 3Galα1,4Galβ1,4Glcβ1,1-Cer
Gb5	globopentaosyl-Cer: Galβ1,3GalNAcβ1,3Galα1, 4Galβ1,4Glcβ1,1-Cer
Blood group H	type 1 sugar: Fucα1,2Galβ1,3GlcNAc
Blood group B	type 1 carbohydrate: Galα1,3(Fucα1,2)Galβ1, 3GlcNAc
Blood group A type 1 carbohydrate	GalNAcα1,3(Fucα1,2)-Galβ1,3GlcNAc
IGLad	Intestinal neutral glycosphingolipid
neolactotetraosyl-Cer	Galβ1-4GlcNAcβ1-3Galβ1-4Glcβ1-Cer
Blood group type O/H type 1 pentaglycosylceramide	Fucα2Galβ3GlcNAc-β3Galβ4Glcβ1Cer
Gb3Cer	Galα1,4Galβ1,4Glcβ1,1-Cer
Gb4Cer	GalNAcβ1,3Galα1, 4Galβ1,4Glcβ1,1-Cer
Forssman antigen	GalNAcα1,3GalNAcβ1, 3Galα1,4Galβ1,4Glcβ1,1-Cer
Paragloboside	Galβ1,4GlcNAcβ1,Galβ1,4Glc-Cer
GM1a	Neu5Acα2,3(Galβ1,3GalNAcβ1,4)Galβ1,4Glcβ1-Cer
GM1	Galβ1,3GalNAcβ1(NeuAcα2,3)4Galβ1,4Glc-Cer
Sialyl-dimeric-Le^X	NeuAcα3Galβ4(Fucα3)GlcNAcβ3Galβ4(Fucα3) GlcNAcβ3Galβ4Glcβ1-Cer
Sialyl-neolactohexosyl-Cer	NeuAcα3Galβ4GlcNAcβ 3Galβ4GlcNAcβ 3Galβ4Glcβ1-Cer
LacNAc (CD75)	Galβ1,4GlcNAcβ1-R
CD75s (α2,6-sialyl LacNAc)	Neu5Ac α2,6Galβ1,4GlcNAc β1-R

Pig Blood Group A Type

A type 1 hexaglycosyl-Cer	GalNAcα1,3(Fucα1,2) Galβ1,3GlcNAcβ1,3Galβ1,4Glcβ1-Cer
A type 4 heptaglycosyl-Cere	GalNAcα1,3(Fucα1,2)Galβ1,3GalNAcβ1,3Galα1, 4Galβ1,4Glcβ1-Cer
A type 1 octaglycosyl-Cer	GalNAcα1,3(Fucα1,2)Galβ1,3GlcNAcβ1,3Galβ1, 3GlcNAcβ1,3Galβ1,4Glcβ1-Cer
A type 1 nonaglycosyl-Cer	GalNAcα1,3(Fucα1,2)Galβ1,3GalNAcα1,3-(Fucα1,2)Galβ1,3GlcNAcβ1,3Galβ1,4Glcβ1-Cer
H type 1 pentaglycosyl-Cer	Fucα2Galβ3GlcNAc-β3Galβ4Glcβ1-Cer
Repetitive A type 1 nonaglycosyl-Cer	GalNAcα1,3(Fucα1,2)Galβ1,3GalNAcα1,3-(Fucα1,2)Galβ1,3GlcNAcβ1,3Galβ1,4Glcβ1-Cer

Chapter 1
Glycosphingolipids (GSLs)

1.1 History and Definition of Sphingolipids and GSLs

Structural lipids in membranes are composed of three general types of membrane lipids: (glycero)phospholipids, sphingolipids, and sterols. Sphingolipids are the derivatives of sphingosine and sites of biological recognition. Sphingolipid does not contain glycerol, different from (glycero)phospholipid. Sphingolipid contains a long-chain fatty acid and sphingosine (4-sphingenine)-based long-chain amino alcohol, and in certain case, phosphoric acid is diester-linked to the polar head group. Sphingolipid has a three subclasses of ceramide derivatives, including sphingomyelins, neural (uncharged) glycolipid, and ganglioside, depending on head groups. Sphingomyelin has phosphocholine or phosphoethanol-amine as a polar head and thus these can belong to a phospholipid. Neutral glycolipid and ganglioside have one or more sugar residues in their head groups and linked C-1 OH of ceramide without phosphate. These sugar-containing sphingolipids are also called GSL.

GSLs are structurally ceramide-bound carbohydrates and fundamentally synthesized in vertebrates and insects, but not in plants [1]. Simply, GSLs are amphiphatic sphingosine-containing glycolipids, where ceramides composed of sphingosines and fatty acids are hydrophobic and carbohydrates are hydrophilic. Thus, GSLs are representatively integral portions with phospholipids and cholesterol in plasma membrane bilayers. The glycolipid is glycans plus lipids in its terminology. However, glycans and lipids are mostly diverse in their structural basis with tremendous derivatives. Therefore, the glycolipids indicate a diverse and broad class of cellular membrane lipids from prokaryotes to eukaryotes. The chemical organization, the International Union of Pure and Applied Chemistry, named the IUPAC, designates glycolipids to compounds carrying a glycan unit glycosidic-linked to a lipid moiety. Although plants and bacterial glycolipids indicate glycoglycerolipids as the major species of glycolipids, vertebrates and insect glycolipids are GSLs backboned by ceramides as glycans carrier. GSLs are structurally diverse, as originated from

C.-H. Kim, *Glycosphingolipids Signaling*,
https://doi.org/10.1007/978-981-15-5807-8_1

sphingolipids [2]. Compared to other known macromolecules as life chains found in cells, information and studies of GSLs are quite lately progressed in their depth and quantities, making them still mysterious and curious. The initial term “Sphingosin” indicates such word as sphingolipids, which were initially named by the German biochemist Johann Ludwig Wilhelm Thudichum who denoted the term it from the enigmatic ancient Egyptian “Sphinx-like” characters due to the Sphinx’s diverse faces and ability [3] in 1884 [4].

The philosophical deduction of Greek Sphinx is considered to be derived from their enigmatically diverse function [5]. Sphingolipids function in cooperative actions with cellular regulatory proteins. Although sphingolipids belong to complex GSLs (http://www.sphingomap.org), sphingolipids include a large group of lipids containing basically large sphingoid bases such as dihydrosphingosine and sphingosine as membrane components and biologically functional actors. The chemically *N*-acylated sphingosine is called “ceramide (Cer),” and Cer is the fundamental precursor of the GSL and sphingomyelin (SM) during their biosynthesis. The known membrane sphingolipids are currently Cer, Cer-a-phosphate, sphingosine (SP), and SP-1-phosphate for the forms of simple structure.

In 1884 J. L. W. Thudichum discovered colorful sphingosine; now sphingolipids are featured with the common structures of sphingoid bases or long-chain bases. The SP has the chemically (2S,3R,4E)-2-amino-octadec-4-ene-1,3-diol structure. For other name, SP is also called sphingoid base or (E)-sphing-4-enine. Therefore, ceramide is the sphingoid base amide-fatty acid-linked form. Ceramides can enzymatically be converted to complex sphingolipid forms, which include complex GSLs, SM, GlcCer, and GalCer. For a minor form, lyso-sphingolipids are also formed through sphingoid bases linked with N-acyl-derivatives-deficient head group. They are lyso-GSLs, sphingosine-1-phosphocholine, and sphingosine-1-phosphate. The GSL nomenclature has been made from IUPAC-IUB guideline for systemic nomenclature of GSLs of the official generalization of the IUPAC-IUB Commission on Biochemical Nomenclature reported in an article reference of the CBN for lipids [6] (Fig. 1.1).

1.2 Classification and Chemical Basis of GSLs

Structures, classification, and nomenclature of GSLs are defined by IUPAC. Monoglycosylceramides are named cerebrosides including GalCer and GlcCer. In gangliotetraosylceramide, “Ganglio” is expressed as Gg. In a structure of Galβ1 → 3GalNAcβ1 → 4Galβ1 → 4Glc, the number of monosaccharides is expressed as a trivial name of “tetra” and a symbol of “Ose4.” In the “Iso (symbol of I)” meaning in the globotriaosylceramide and isoglobotriaosylceramide structures, the difference between the second monosugar and the third monosugar linkage (glycosidic linkage α1-4 vs. α1-3) leads to the terminology described as the “Iso.” Seemingly, the term “Neo”(n) implies the linkage difference between lactotetraosyl-

Johann Thudichum,
1829–1901

Fig. 1.1 Johann Ludwig Wilhelm Thudichum also known as John Louis William Thudichum (August 27, 1829–September 7, 1901), a German-born physician and biochemist. Adapted from en.wikipedia.org/wiki/Johann_Ludwig_Wilhelm_Thudichum

Table 1.1 Trivial name, symbol, and structure of GSLs

Trivial name	Symbol	Structure
Gangliotetraosylceramide (Asialo GM1, Gg4)	GgOse4Cer	Galβ1 → 3GalNAcβ1 → 4Galβ1 → 4GlcCer
Globotriaosylceramide (Gb3)	GbOse3Cer	Galα1 → 4Galβ1 → 4GlcCer
Isoglobotriaosylceramide (iGb3)	iGbOse3Cer	Galα1 → 3Galβ1 → 4GlcCer
Lactotetraosylceramide	LcOse4Cer	Galβ1 → 3GlcNAcβ1 → 3Galβ1 → 4GlcCer
Neolactosylceramide	nLcOse4Cer	Galβ1 → 4GlcNAcβ1 → 3Galβ1 → 4GlcCer

Cer and neolactosyl-Cer (glycosidic linkages of β1-3 and β1-4) in the third and fourth monosugars (http://identifiers.org/lipidbank) (Table 1.1) [7].

The terminological meaning of GSLs has been expanded and defined to multiple components that consist at least of one sphingoid and one monosaccharide residue. The GSLs are divided into many categories. Monosaccharide β-linked Cer group includes Glc for GlcCer or Gal for GalCer and can be further converted to sulfated GSLs like sulfatides. GSLs are further subgrouped into two groups of neutral or acidic forms (gangliosides and sulfatides). Acidic GSLs consist of two similar families of the sialosyl-GSLs, which is termed gangliosides, and the sulfo-GSLs, which is referred to as sulfatides [8]. In higher vertebrates like mammals, GSLs are accurately subdivided into four main groups of lacto (Lac)- (type 1), neoLac- (type 2), globo-, and ganglio-series GSLs, depending on their structure basis. Similarly, Galβ1, 4Glcβ1Cer, or LacCer is further branched to generate globo-, isoglobo-, lacto-, neolacto-, and ganglio-series. Neutral glycolipid has 1–6 sugar residues (sometimes 6 more) including N-acetyl-D-galactosamine (GalNAc), D-Gal, and D-Glc. Cerebroside is the one sugar residue attached to ceramide.

The ganglio-series are generated through SA addition. Gangliosides are also classified to the ganglio-, lacto-, and globo-series of GSLs [9]. Among them, ganglio-series, later termed gangliosides, carries one or more sialic acid (SA) residues commonly found on cell surfaces. GSLs are also generated for the compliant similarity through different combinations such as globo- versus isoglobo- or lacto- versus neolacto- or ganglio-series of GM1a and GM1b. Among gangliosides, the simplest form is GM3 that carries α2,3-sialyllactose (SL) in carbohydrate part. GSLs are differently named by conventional nomenclature systems, and most GSLs are used by their historic terminologies in the cases of gangliosides. For example, the GM1a and GM1b gangliosides have their specific structures of the Neu5Acα2,3(Galβ1,3GalNAcβ1,4)Galβ1,4Glcβ-1-Cer (d18:1/C18:0) and Neu5Acα2,3Galβ1,3GalNacβ1,4Galβ1,4Glcβ-1-Cer, respectively, which have commonly the Cer- chemical structure. They are also termed with the Ganglio common structures. The Ganglio series terminology uses the linkage location and position of the SA (NeuAc) with the chain through Arabic superscripts and Roman numerals to indicate the NeuAc-linked OH position. Thus, GM1a is indeed II3-α-Neu5NAc-Gg4Cer, reading to II3-α-N-acetyl-neuraminosyl-ganglio-tetraosyl-Cer. Similarly, GM1b can be expressed as IV3αNeu5-NAc-Gg4Cer-, reading to IV3-α-N-acetyl-neuraminosyl-gangliotetraosyl-Cer. For the additional modifications like 9-O-acetylation of SA or intramolecular lactone formation, the description is added to the term. Certain glycans are still called by conventional and historical names for the case of Lewis blood antigens. Chemically, GM3 contains only a trisaccharide with terminally sialylated lactose (Lac) moiety in SA α2-3-linked Gal. GM3 is used as the core precursor in glycan biosynthesis of the gangliosides.

Chemically, GSLs are regarded to be the most “un-lipid-type” and therefore, not easy to isolate, purify, and analyze from cell membranes. Traditionally, there are two different methods to isolate the GSLs. Typically, the GSLs extraction method is applied using the Folch extraction [10] and the di-isopropyl ether/1-butanol extraction [11], when tissue lipids are subjected to isolate. In the methods, the gangliosides are fractionated into the aqueous layer due to the hydrophilic characters. In contrast, it is noted that other lipid fractions are typically fractionated into the chloroform-extracted layer or di-isopropyl ether/1-butanol-extracted layer due to their hydrophobic/lipophilic properties. For purification of gangliosides, two different methods of solvent partition (1-butanol/di-isopropyl ether/50 mM NaCl solution and Folch partition) and ion exchange chromatography [10, 11] are applied. Among them, the 1-butanol/di-isopropyl ether/50 mM NaCl partition solution is convenient for total cell lipid extracts. For desalting, Sep-Pak C18 cartridge kits are currently applied from commercial sources; however, another method, gel filtration using Sephadex G-50 column, is rather inconvenient for di-isopropyl ether/1-butanol/50 mM NaCl preparation [12]. Neutral GSLs as non-sialyl GSLs are also isolated by acetylation [13] or non-acetylation [14].

1.3 Biological Aspects of GSL Glycan-Receptor Interaction

GSLs are also essential for the correct functioning of cell surface receptors. Gangliosides are present in substantial amounts in cell membranes and together with globosides are present in the blood and myeloid cell membranes. Gangliosides plus the lacto- and neolacto-series GSLs are involved in cell recognition like blood group determinants. Glycolipids on cells determine their antigenicity and their status of differentiation, undifferentiation like embryonic status, normal and malignant status. Gangliosides are investigated as potential antitumor vaccines. The GM1 has effects on nerve growth stimulation and experimental Parkinsonism protection. The sphingolipids can bind to proteins to exert their unusual functions on the cellular membranes. For example, such binding of sphingolipids and proteins is involved in various cellular events known in apoptosis, cell survival, and aging [6]. In the glycan-linked Cer, glycan moieties of GSLs function as a biological serving cooperator. The biophysical functions are relied on two different theoretical roles displayed in membranes. GSLs spontaneously organize microdomains called "lipid rafts" in plasma membranes (PM) to form functionally active "glycosynapses" with related decorating-receptor proteins or ion channels. In one side of the GSLs, the carbohydrate glycan chains in GSLs can bind their specific receptors. For example, in the class of protein-glycan bindings, it is a type of the protein-carbohydrate interaction (PCI). Also, in the case of the cognate glycan-glycan binding, its interaction and recognition are called carbohydrate-carbohydrate interaction (CCI). Actually, upon stimulation of cells, GSLs spontaneously construct lipid rafts in biological membranes to form so-called glycosynapses with receptor proteins. Resultantly, the glycan moieties in GSLs interact with receptors to afford endocytosis, vesicular transport trafficking, lipid rafts formation, intracellular signaling, and cell-cell interaction [15]. For more evolved types of GSLs, gangliosides as the SA-carrying GSLs are frequently engaged in the formation of microdomain lipid rafts and regulate cellular function through activation or signal transduction. With regard to functional roles, GSLs on cell membranes are regarded as neuronal or embryonal development biomarker molecules as well as pathogenic toxin and entry receptors due to their biological functions in mammals [16, 17]. Most representatively, a ganglioside GM1 is a GSL with one SA attached to ceramide as a membrane microdomain component named lipid rafts.

GSLs influence disease status, where apart from their actions to normal cell behavior, GSLs patterns, quantities, and distribution greatly influence pathologic processes by dysfunctional pathologic conditions. Such dramatic examples are seen by inherited metabolic disorders, neuronal diseases, autoimmune diseases, and cancers. In addition, upon infections, GSL glycans function as attachment and entry sites for infectious viruses, invasive bacteria, and bacteria-produced exotoxins. In the side of structural importance and significance of the cellular membrane GSLs organization, currently accumulated results suggest that GSLs diversity and heterogeneity are directly related to the formation of the ordered membrane protein compartments in many different phenotypes of mammal cells. As GSLs are comprised of a glycan carbohydrate structure linked to sphingolipid ceramide in tails, GSLs are therefore amphipathic lipids which consist of carbohydrate moieties bound to ceramides. At the cellular level, GSLs are differentially expressed in a stimulus-depending mechanism. The carbohydrate structure, distribution, quantity, and composition also influence on the ordering of GSLs and surrounded phospholipids in plasma membrane via mosaic theory. Such environmental situation also leads to the better condition toward the extension and conformation of the GSL carbohydrates in the membranes. With the situation in carbohydrate size and qualified structure between the phospholipid headgroups and GSL components, the surrounded lipids are naturally ordered.

1.4 Glycan-Receptor Interaction Aspects of Pathogenic Infection and Toxin of GSLs

Diverse structures of GSLs influence receptor signal signaling by interacting with specific receptor regions, although the biological and evolutionary aspects of the evolution and diversity of the GSL glycans are largely unknown. They are generally known to function as antigens and also receptors for bacterial toxins. In cell-cell interaction, they are directly involved in recognition, adhesion, and signaling in target recognition events or initiation and modulation of cellular responses. The glycans on GSLs can recognize, bind, and interact with many membrane receptors [18–20], where GSLs and cholesterol are associated with the cellular membrane to associate microdomain lipid rafts and function as a signaling complex. For example, the well-explained interaction of the GSLs, which potentiates intracellular signaling pathways, is observed in infectious diseases and mammal immunity [21, 22], as GSLs recruit accessary and adaptor proteins such as caveolin to form microdomains for signaling cascades [23, 24]. Such functions are basically derived from the hydrophobic properties in GSL structures [2].

During viral infection of host cells, viruses utilize surface molecules and their specific receptors on the host cells to penetrate, enter, infect, and replicate during life cycle. Even in the same virus family, each virus member differentially targets different cells or different tissues in the same hosts. This is simply due to different receptor specificity. Microbes utilize the different glycan structures expressed on each different cell surface to take advantage of recognition and attachment during infection. Because virus receptor recognition is the subject of variation in interaction with SAs and the binding site on the capsid proteins is variable, several factors including receptor structure, host preference, and viral evolution in gene level are fundamental aspects of the infectious diseases. For specific viruses like the polyomaviruses, non-enveloped viruses including rota-, calici-, parvo-, and polyomavirus families, ganglioside binding defines the internalization pathway into host cells, indicating that GSLs are receptors for the non-enveloped virus families. But in other viruses, the microbe and carbohydrate recognition and interaction open a host susceptibility as a critical determinant. Hydrophilic headgroups in membrane GSLs interact with virus receptors and bacterial virulence factors to initially function in the environmental formation of biological societies [25–38].

Ganglio- and globo-series of GSLs independently function for each ligand during viral and bacterial infections. Glycan moieties in GSLs are recognized with viral proteins, or they interact with membrane molecules or certain carbohydrates present on the PM surfaces of themselves or neighboring adjacent cells, depending on the GSLs-glycan moieties composition and structure [18, 22]. For the molecular interaction, the surfaced viral proteins generally recognize SAs located to each terminal glycans in most sialic acid-utilizing viruses. In addition, changes in binding-pocket amino acids also regulate the binding capacity of sialylglycan receptors [39]. Therefore, comprehensiveness, complexity, and flexibleness of molecular interaction in virus-glycan receptors contribute to the virus-host specificity and host-targeted evolution of virus.

Because the glycan structure and distribution of the GSLs recognize at the interface on proteins [18–22, 40], the question how cell surface receptors can be rested or activated for signaling behaviors upon their interaction with GSL is raised. GSLs interface with counterpart protein domains located in signaling molecules that as such GSL-sensing molecules. The GSL-mediated signaling is straightforward due to their direct ligand-receptor axis, as the ligand-receptor bindings are not interfered by other GSLs. This straightforward property of GSLs can open a new medium for the signaling regulation in development. Therefore, such functions of GSLs can also influence progenitor cell differentiation and morphogenesis during embryonic or neuronal development. GSL-protein interactions are quite attractive for the life science [40]. Protein-carbohydrate interaction (PCI) process is generally dependent on multivalent ligand chemistry, as it is compared to the low affinity ligand chemistry of CCI [41]. GSLs sugar moieties also potentiate CCI [42], PCI [43], or protein-sphingolipid interaction (PSI) [6, 43, 44] for signaling receptors [22]. The chemical strength of PCI is chemically weak due to multivalent bonds-based low affinity [45], but crucial for microenvironmental biological events such as temporary cell adhesion or leukocyte homing. The carbohydrate moiety of membrane glycoproteins and glycolipids receptors interact with their ligands in the fashion of PCI and CCI [46]. Although PCIs are in low affinity binding, the case of CCIs between GSLs is much low or ultra-low affinity, but with some advantages during the initial signaling events such as cell adhesion stage of development [47]. For experimental CCI management, multivalent glycan analysis is rather thought to be crucial due to accumulated total interactions [48]. Typically, membrane microdomain lipid rafts exhibit such multivalent minor bonds like CCIs. As typical CCI types, Le^x/Le^x interactions in mouse embryogenesis occur with multivalent minor, but multiple interactions [49]. In another CCI case, surfaced GM3 on metastatic melanoma cells interacts with gangliotriaosylceramide (Gg3) on lymphoma cells at the initial adhesion stage of interaction between melanoma cells and the endothelium [50]. Currently, the CCI process has been limitedly reported at the experimental level [50, 51]. In addition, the GSL-protein interactions are not well understood with the detailed interaction mechanism [43], although only exceptional cases for the

GSLs-protein receptor interaction have been reported [22, 52, 53]. Each glycan composition of GSLs is specific for each specific signaling in the ligand-receptor interaction. As a CCI type, the GM3 and lactose interaction between glycolipid Langmuir monolayers and glycol-dendrimers were previously reported using the glycol-micelles [51]. The GM3 and Gg3 interaction pair has been traced using different GSLs or Gg3-carrying-polymers [54]. GSLs metabolism and signaling are the recent leading field for apoptosis, cancer, cell survival, and aging by PSI. Examples of the GSLs-binding proteins are known for receptor, effector, enzyme, and transporter [6]. Infection events highlighted by various virus and bacteria are mediated by GSLs-carrying glycans as attachment sites also for primary targets for bacterial toxins.

Currently in understanding of the GSLs, several basic questions are raised in the present review: How do carbohydrate glycans recognize, bind, interact, friendly molecules-recruit, and transduce their signals to their counterparts to evidence their roles? If glycolipids, how the GSLs form their lipid rafts or microdomain on the cellular plasma membrane (PM)? How such GSLs, SMs, and Cers are classified during biosynthesis of GSLs in membrane-bound organelle and sugar nucleotide? Therefore, in this review, the PCI, CCI, or PSI signaling is discussed with an extensive introduction to GSLs-receptor protein interaction to broaden pathological manifests of GSLs in cellular and molecular interactions.

Chapter 2
Mammal GSL Synthesis Via ER and Golgi Network

2.1 Synthesis of Ceramide and Simple Derivatives

GSL synthesis and metabolism are carried out via endosomal organelle of endoplasmic reticulum (ER) and Golgi systemic network, in the secretory organelle membrane system [16]. As a starting lipid structure to generate GSLs, the Cer is used. As described earlier, the aminoalcohol sphingosine with long chain is a key constituent of Cer [1]. Sphingosine is synthesized from serine and palmitoyl-CoA with fatty acid-amide linkage because the GSL biosynthesis is initiated by the serine palmitoyltransferase, an amino acid-fatty acid ligase. The serine palmitoyltransferase catalyzes the synthesis of 3-keto-sphinganine from palmitoyl-CoA and serine. The keto group is further reduced to the hydroxyl group to synthesize the sphinganine named dihydrosphingosine and the C-4 position of the formed dihydrosphingosine is further hydroxylated to make the phytosphingosine (4-hydroxysphinganine). Dihydrosphingosine and phytosphingosine are then converted to dihydroceramide or phytoceramide, respectively, by each substrate-specific acyl transferase. Ceramide is converted from dihydroceramide by a desaturase, which is the enzyme to form a double bond at position C-4 of a precursor dihydroceramide [55].

Structurally, the Cer is diverse by the different chain lengths of acyl groups [2]. Monoglycosylceramides or oligoglycosylceramides are glycosidically linked to the Cer. Galactosyl- or glucosylceramides (GlcCers) are the most basic oligoglycosylceramides. The globo-structure has longer oligosaccharide chains in ceramides. GSLs are further classified into acidic GSLs and neutral GSLs by the charged glycans of SA, where the SA is largely a type of *N*-glycolylneuraminic acid (NeuGc) or *N*-acetylneuraminic acid (NeuAc). Exceptional acidic GSLs included 2-aminoethyl-hydroxyphosphoryl groups, phosphate monoester group, phosphate diester group, or uronic acid residue. The carbohydrate epitope of CD57 or human natural killer cell-1 (HNK-1) antigen is 3′-sulfated glucuronic acid (GlcA), and this carbohydrate is linked to an *N*-acetyllactosamine (LacNAc) on glycoproteins and

C.-H. Kim, *Glycosphingolipids Signaling*,
https://doi.org/10.1007/978-981-15-5807-8_2

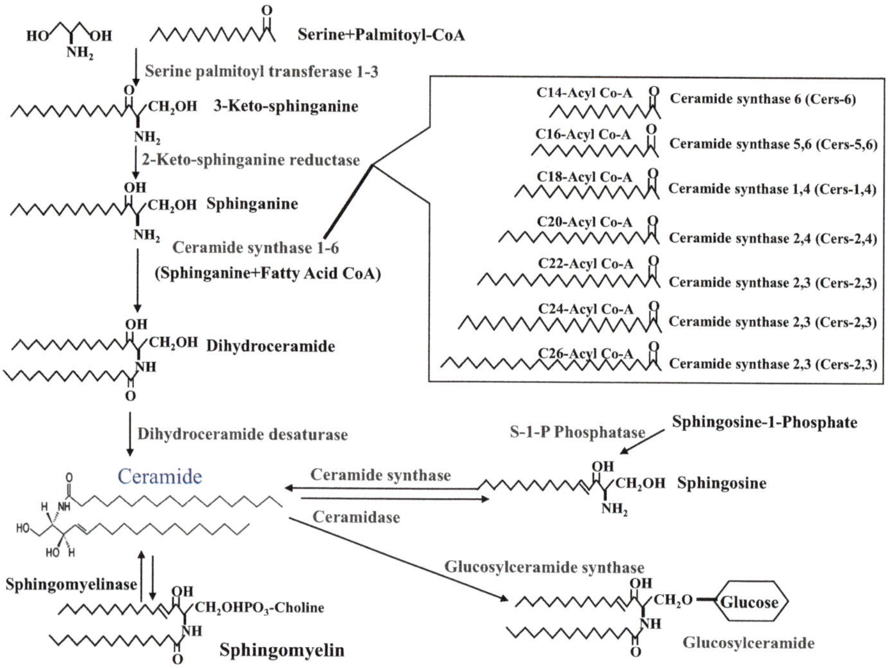

Fig. 2.1 Biosynthesis and metabolism of the simple GSLs

glycolipids [56]. Interleukin 6 (IL-6) has a lectin-like activity for carbohydrate ligands on the HNK-1 epitope 3-sulfated GlcA on certain mammalian N-glycans among eukaryotes. The gp130 has the HNK-1 epitope. More specifically, HNK-1 or CD57 has a structure of 3-O-sulfo-GlcAβ1,3Galβ1,4GlcNAcβ1-R. CD57 carbohydrate glycan is the HNK-1 epitope. It contains a 3′-sulfated GlcA attached to LacNAc of N-, O-glycans and glycolipids [57]. CD57 carbohydrate directly associates with the selectins of CD62L, which is called the L-selectin, LECAM-1 or Mel-14, and CD62P, which is called the P-selectin, GMP-140, LECAM-3 or PADGEM. CD57 leads to the development and formation of the mammal nervous system and synaptic plasticity.

The GSL biosynthesis is initiated through condensation of a sphingolipid base, sphinganine, or sphingosine (SP) with an acyl-CoA to generate Cer in the ER [58] (Fig. 2.1). Cer is therefore the most basic and starting precursor of **SM** and GSLs in mammals. As a key enzyme of Cer synthesis, Cer synthase (CERS) is the primary enzyme of ceramide synthesis. The fatty acid chains in GSLs are generally longer than that in SM. The Cer is de novo generated in the biosynthetic pathway of ER and continuously delivered to the Golgi system named Golgi network by specific trafficking pathways. Then, SM and GlcCer are biosynthesized as the first step of glycosylation. In mammals, dihydroceramide or Cer is important for the PM fluidity and plasticity for functional expression of the cells. If the synthesis of the dihydroceramide or Cer is defected, the fate of the cells is not established. As reflected from the significance of Cer or dihydroceramide, the synthetic enzymes of CERS are multiply located on the genetic background. Totally, six kinds of CERS are known for the de novo dihydrosphingosine *N*-acylation to finally synthesize dihydroceramide species. Moreover, another synthetic pathway, named salvage pathway, leads to *N*-acylation of sphingosine to form ceramides [59]. Each CERS prefers for each specific acyl-CoA for the (dihydro)Cer synthesis. Thus, the regulated molecular ratio of CERSs implicates for the fatty acid composition and distribution of sphingolipids [60]. The chain length of acyl group in sphingolipids is responsible for their intracellular delivery, endocytosed trafficking of SM class, and recycling to the PM by the transportation pathway via ER-Golgi apparatus network [61]. Despite the current progress in Cer synthesis, the questions on how the fatty acid acyl chain length in sphingolipids influences on cell fate and cell function, signaling, or CERT-independent Cer transport are poorly understood, requiring the future progress in the fundamental understanding for the goals.

The monoglucosyl derivatives are biosynthesized by the enzymatic transfer of the Gal or Glc moiety from donor substrates of sugar nucleotides such as UDP-Gal or UDP-Glc to the Cer in a mode of O-glycan biosynthesis [62]. In the lumenal surface of the ER, the Cer is galactosylated to generate galactosylceramide (GalCer) [63], by UDP-galactosyl:Cer galactosyltransferase. GalCer species are transported through the ER to form 3'-*O*-sulfate-containing form of sulfatide (sulfogalactosyl ceramide) and Gala-series (galabiosyl ceramide) GSLs of long glycan chains by glycosyltransferases [64]. Also, GalCer is directly transported to the Golgi complex to form sulfogalactolipids, which are necessary in the brain because they are mainly composed of the myelin sheath in neuronal cell to form axon synapse. In some cases, GalCer is directly sialylated to form the sialosylgalactosylceramide ganglioside GM4 [NeuAcα2–3Galβ1–1′Cer] as a myelin sheath component [16]. The GM4 is known to express predominantly in the brain and erythrocytes of mammals through the GalCer sialylation. Whereas LacCer is sialylated, GM3 is generated. GM4 is a rarely synthesized ganglioside because it is mainly expressed in the white matter of the brain as a minor ganglioside, and GM4 was specifically found in human astrocytes and brain myelin region [65]. GM4 is less polar ganglioside and expressed in various vertebrates. In mammals, although most ganglio-series GSLs are synthesized from the simple GM3 by GM3 synthase (ST3GalV/SAT-1) as a starting enzyme for ganglioside synthesis [66], GalCer, but not GlcCer, is the acceptor substrate of GM4 synthase. However, the function of GM4 synthase is molecularly not defined yet.

On the other hand, during the SM synthesis, Cer is trafficked from ER to the trans-Golgi region of the Golgi complex located at the endosomal membrane interaction sides. Also, during metabolism, catabolism, and function, the lipid transfer between different membranes of different endosomal organelles is appropriately carried out. Certain proteins are involved in catalytic translocation of lipids through the intermembranes and the proteins are called lipid transfer proteins (LTPs). In fact, mammal organisms have various LTPs because LTPs are essential for the appropriate transport of lipids between organelles at organelle membrane interaction sites. Ceramide transport protein (CERT) is a representative example [67]. SM synthase-1 de novo synthesizes the SM in the medial/trans-Golgi network, whereas UDP-glucose Cer glucosyltransferase (UGCG) as the GlcCer synthase (GCS) is located on the cis/medial Golgi network [68]. Therefore, most GSL glycosyl transferases are membrane-specific, while only GCS is cytosolically localized. The transport of Cer is carried out via intracellular transport vesicles or by the CERT in a fashion that CERT picks up Cer from ER membranes to deliver to the trafficking machinery of trans-Golgi network (TGN) upon the SM synthesis beginning [69] (Fig. 2.1). Although there are two different ways in the ER-to-Golgi translocation of Cer, the CERT-dependent and CERT-independent transport are reported [70, 71]. Only CERT-mediated non-vesicular Cer species is transported from the organelle ER/Golgi trans-network as the major and acting pathway for the SM biosynthesis [72]. Cer transported to Golgi system is further glycosylated by various glycosyltransferases localized on a long Golgi *cisternae*. Cer, or the nonreducing end of glycosylated Cer, is used as acceptor substrate [73].

GlcCer, the simplest GSL synthesized by the added glucose residue to Cer, is synthesized on the cytosol area near to the early stage of Golgi-vesicle membranes by the enzymatic catalysis of GlcCer synthase working in the cellular cytosolic fraction. Most synthesized GlcCer are therefore localized on the cytosol leaflet side of *cis*-Golgi network apparatus membranes [69]. More specifically, the other glycosyltransferases are located at the regions of carbohydrate extension, localizing as organellar membrane-anchored proteins faced with the Golgi apparatus lumen. However, the exceptional GalCer synthase is located within the ER lumen. Thus, GlcCer is biosynthesized from the Cer moieties cytosolically embedded in the Golgi membrane surfaces. Like Cer, GlcCer can then be transported using two independent pathways to the lumen leaflet of the membrane in the Golgi cisternae complex or the Golgi apparatus complex by means of membrane vesicular trafficking process. Another pathway is evolved to the biosynthesis of the SA-lacking type of globo-series of GSLs, where the translocation is carried out by the TGN-released lipid transfer protein FAPP2 through picking up GlcCer, which is localized in the cis-Golgi membrane region, to the lumen side [74]. The early Golgi system-transported GlcCer is oriented to the lumen side of the Golgi system in which the orientation is catalyzed by a flippase enzyme (Fig. 2.1). GlcCer-based glycan size is extended by various glycosyltransferases to form the complex forms of GSLs in a mode of Golgi apparatus-based *N*-glycan synthesis [75]. Therefore, GlcCer located on the side of cytoplasmic membrane is delivered from the *cis*-Golgi region to the *trans*-Golgi region by a trafficking protein FAPP2. However, the subsequent GlcCer flipping process to the Golgi lumen is not clearly explained yet. When GlcCer is once delivered to the luminal Golgi and TGN leaflet, further progressing of GlcCer is continued on to form Gal-GlcCer named **LacCer** by galactosyltransferase which transfers Gal residue. Once synthesized LacCer is not reversely delivered back to the membrane leaflet sides of cytosol. LacCer is therefore utilized for straightforward synthesis of all downstream GSLs [16, 76]. The biosynthetic pathway of the complex GSLs is featured in Fig. 2.1. From the TGN, the freshly formed GSLs and SM are trafficked and delivered to the PM by endosomal vesicular transporters, allowing further modification by membrane-bound glycosidases. These membrane-specific modification events regulate changes in GSL distribution and composition at the cell PM, allowing membrane GSLs internalization to the endosomal and lysosomal organelle by helps of each hydrolase.

Apart from the GlcCer transport, the ABCB1 known as the MDR1 drug efflux pump is known to flip the GlcCer. However, the similar LacCer or its analogs cannot across the ABCB [77]. Because MDR1 is expressed in Golgi, MDR1 inhibitors can prevent cellular GlcCer levels and neutral GSLs. A representative MDR1 inhibitor, cyclosporin, also decreases the Gb_3 levels in the Fabry mouse [78]. However, MDR1 inhibition does not decrease in ganglioside biosynthesis, despite the effective decrease in GlcCer and LacCer levels, indicating the existence of alternative pools for gangliosides and neutral GSLs synthesis [79].

2.2 Synthesis of Complex GSLs Including Lacto, Globo, Asialo, and Ganglio Series

Diversely synthesized and distributed GSLs are subclassed, depending on their charged molecules (acidic, neutral, basic) or carbohydrate chain structures shown in the known glycans such as lacto-GSL, gala/neolacto-GSL, ganglio-GSL, and globo-series GSL. Acidic GSLs contain negative charged monosaccharides such as SA or sulfate groups in carbohydrate residues, while basic GSLs are not frequently found in their structures of the membrane lipids (Fig. 2.2). In the charged GSLs, they are further subclassified into acidic (negative charge) and neutral non-acid groups. In addition, the acidic GLSs are further re-subclassified into esterified sulfate-conjugated sulfatides and gangliosides, SA-containing GSLs. For another classification, GSLs are also subjected to subclassification from the internal glycan core structures. Type 1 GSL or lacto series GSL, having the structure of Galβ1-3GlcNAcβ1,3Galβ1,4GlcβCer, type 2 GSL or neolacto-series/type GSL having the structure of Galβ1,4GlcNAcβ1,3Galβ1,4GlcβCer, and type 4 GSL or globo-series GSL, having the structure of Galα1,4Galβ1,4GlcβCer core chains, are originated from the LacCer as a precursor, and they are the common non-acidic or neutral GSLs. In contrast, gangliosides are mainly originated from ganglio-GSL structure of Gal-[Neu5Acα2–3]-β1-4Glc for GM3 or neolacto core carbohydrate structures. The neolacto and lacto core structures have frequently been found in glycoprotein carbohydrates. However, the core carbohydrate structures of the globo-series GSL and ganglio-series GSL are only found in GSLs. For the globo-series GSLs, they are synthesized from LacCer (CDw17). Globotriaosylceramide (Gb3, CD77) having the structure of Galα4Galβ4GlcβCer is also designated to be blood group P^k antigen or CD77 antigen. This antigen is synthesized by the Gb3(CD77)-synthesizing enzyme, α1,4Gal-T, using the LacCer as acceptor substrate. The most important aspect of Gb3 is in the regulation of cell functions. Globoside Gb3 was discovered as a P-type blood group-associated antigen and later precisely named P^k antigen. Gb3 is now known as an infection receptor for bacterial Shiga toxins (Stx) [80]. Immature

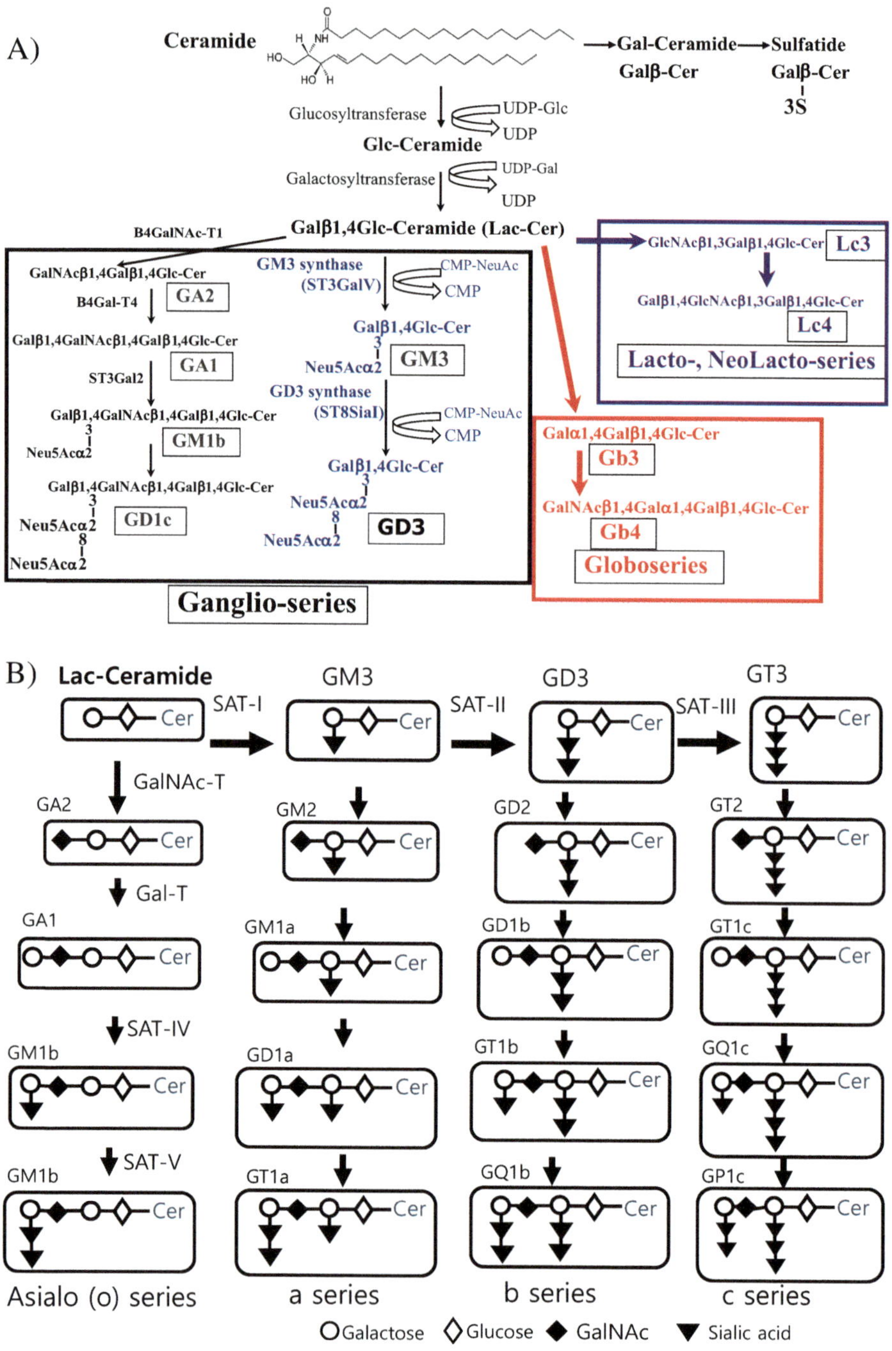

Fig. 2.2 Mammal glycosphingolipid biosynthetic pathway. (**a**) GalCer, GlcCer, LacCer, a and b type gangliosides (blue words), Globoseries and Lacto-series and Neolacto-series. (**b**) Asialo o-series as well as a-, b-, and c-series gangliosides are systematically synthesized from Lac-Cer, GM3, GD3, and GT3

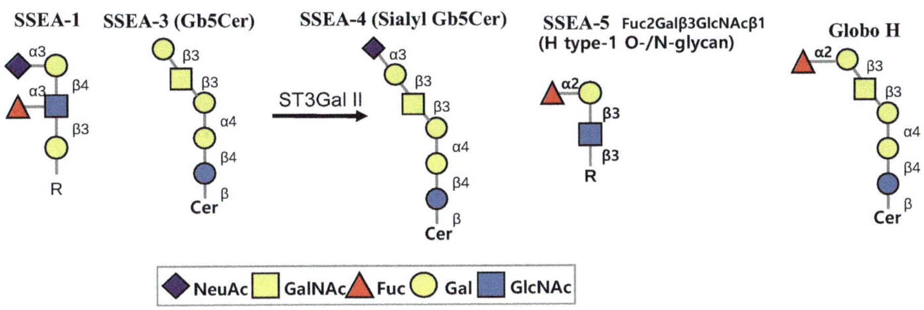

Fig. 2.3 Structures of stage-specific embryonic antigens (SSEA), a class of carbohydrate epitopes stem cell differentiation

B-cells largely express Gb3, and many tumor cells such as Burkitt's B-cell lymphoma also express Gb3 [81, 82]. Gb3 expression is associated with metastatic and invasive potentials in colon cancers, and the lack of Gb3 synthesis contributes to inhibition of tumor cell progression [83]. With regard to the physiological function of Gb3 in cancer progression, tumor surface-specific heat shock protein 70 (HSP70) is reported to be directed by the Gb3 expression [84]. The second role of Gb3 is depicted from the metabolic malfunctional Fabry disease, an X-linked lysosomal disease, which is the phenotype of α-galactosidase A-deficient disease [85]. The α-galactosidase A is the known hydrolytic enzyme for Gb3 in lysosome and its deficiency accumulates Gb3 in lysosome, causing cardiovascular dysfunction. Surfaced Gb3 on host cells also recognizes the HIV gp120 protein to allow HIV attachment, adhesion, entry and penetration as well as the *E. coli* AB5 verotoxin [86].

Globoside Gb4 having the carbohydrate structure of GalNAcβ3Galα4Galβ4GlcβCer is markedly expressed in human erythrocytes [87]. Gb4 is synthesized from globoside Gb3 through catalytic transfer of the β1-3GalNAc residue by specific enzyme β1,3-N-acetylgalactosaminyltransferase enzyme (β1,3GalNac-T) with predominant expression during embryogenesis [88]. Gb4 binds to nLc_4 (Galβ1,4GlcNAcβ1,3Galβ1,4GlcβCer) to induce cell adhesion-related signaling [89] (Fig. 2.2). Globoside Gb5 is a linear type of pentasaccharide having carbohydrate structure of the Glc-Gal-Gal-GalNAc-Gal-Cer, as a Cer-linked form. The Gb5 is a well-known antigen, named stage-specific embryonic antigen (SSEA)-3. Distinct carbohydrate oligomer and surface-specific biomarker, which are specifically synthesized in the embryonic stem (ES) cells of human, are not found in the differentiated cells [90, 91]. If a SA residue is attached to SSEA-3, SSEA-4 is made (Fig. 2.3). GSLs currently include ganglioside, lactoside, neolactoside, globoside, isogloboside, molluside, and arthroside. Their structures and nomenclatures have been shown in Fig. 2.4. In mice, SSEA-4 is a specific stem cell marker that is highly differentiated ectodermal-lined stem cells, although SSEA-4 is expressed in the undifferentiated ES cells [92]. During differentiation of human ES cells, GSLs expressed in the cell surfaces are changed from originally synthesized lacto-GSLs or globo-GSLs to ganglio-series of GSLs. Therefore, ES cell

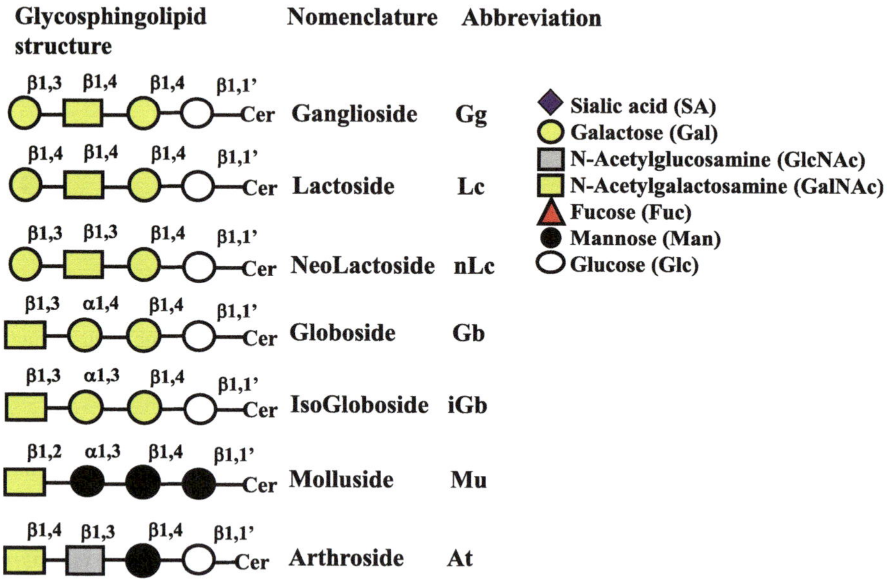

Fig. 2.4 Glycosphingolipid nomenclature

differentiation is associated with highly complex types of GSLs [93]. Regarding globosides and gangliosides, the question is why and how they influence stem cells (SC) differentiation. One possibility is that they function as the cell attachment sites in the glycosynapses formed cells [94]. The biosynthesis of GSLs downstream of LacCer is thought to be regulated by many factors including different glycosyltransferases, lipid transfer proteins, sugar nucleotides, and sugar nucleotide transporters; however, the exact mechanistic explanation of the regulation is unclear [73, 95].

2.3 Biosynthesis and General Aspects of Gangliosides

The gangliosides are the SA-containing GSLs, more specifically, a SA-carrying glycan part and a lipid part as hydrophobic backbone, named Cer. In 1942, a German chemist Ernst Klenk for the first time discovered the gangliosides and gave the terminology of "ganglioside," because gangliosides are mainly expressed and distributed molecules in "Ganglionzellen" (neurons) after their preparation and isolation from brain tissue [96]. The gangliosides were discovered and isolated in the studies of certain neurolipidoses, Niemann-Pick disease, and amaurotic idiocy Tay-Sachs disease. Largely accumulated lipids in the brain ganglial cells were sphingomyelin species in Niemann-Pick diseases. In Tay-Sachs diseases, unknown types of GSL are featured of acidic properties and purple colored during staining with Bial's sugar reagent. These substances were named gangliosides in analogy to

Thudichum's cerebrosides due to the initial location on ganglia cells. For an inaugural article regarding the chemical basis of neuraminic acid (NeuAc) and gangliosides, which Ernst Klenk discovered and isolated, he described the background and history of gangliosides (Chem Phys Lipids. 1970, Vol. 5, No. 1, 193–197). The isolated brain ganglioside preparations were heterogeneous in the structure diversity, evidenced in 1956 by Svennerholm. Gangliosides were further re-classified by Svennerholm and termed, depending on the SA residue number and relative mobilities of SA residues on chromatographic analysis [97].

Compared to glycerolipids embedded into PM, the sphingolipids build on the long-chained amino alcohol SP. The SP is further attached to a fatty acid species through an amino group to consequently yield different Cer species. Three monosaccharides-carrying Cer, named ganglioside GM3, is most abundant and simple in mammals. Their structural derivatives are expressed, depending on growth stages and cell types. To date, 100 more different structures of gangliosides are known from the vertebrates. Each carbohydrate species is classified by ganglioside biosynthetic pathways. The biosynthesis of gangliosides in a sequential order of glycosylations is carried out by two different pathways of "A," synthesizing gangliosides of GD1a, GM1a, and GM2 and another pathway "B," synthesizing gangliosides of GD1, GD2, GD3, GT1b, and GQ1b from a commonly found, simple precursor GM3. The synthesis is carried out in the ER/Golgi system for the most cases. In certain condition, the gangliosides are internalized via endocytosis from the extracellular milieu, although some questions on how these ganglioside molecules are efficiently translocated to the membrane lipid rafts of cells are unclear. Gangliosides belong to a GSL subclass, which distinctly consists of the SA residues. SA residues are linked through SA-α2,3-Gal or SA-α2,6-Gal linkage. Also, a SA-α2,6 attachment is catalyzed to GalNAc and α2,8-SA is linked to α2,3-SA. As earlier described, the predominant SA type is Neuα5Ac in humans, and they are used as cellular membrane receptors for attachment of many infectious viruses [98, 99]. On the other hand, Neuα5Gc is not synthesized in humans due to deletion in exon region of the coding enzyme gene in the CMP-Neu5Ac to CMP-Neu5Gc hydroxylase, which is caused by a species-specific deletion mechanism appeared during evolutional adaptation [100]. Ganglioside synthesis commences with the ER-resident Cer species and the Cer-linked intermediates are sequentially transported to the Golgi complex apparatus. Cer is modified by the enzyme action of a specific UDP-Glc:Cer glucosyl-Transferase (UGCG), to yield the GlcCer. GlcCer is further converted by galactosyltransferase I (GalT-1) to LacCer.

For the simplest form of gangliosides, GM3 is enzymatically formed from its precursor LacCer by the specific enzyme, LacCerα-2,3 sialyltransferase (ST) or GM3 synthase (ST3GalV/SAT-1). This indicates that LacCer is sialylated by GM3S or GM3 synthase to generate GM3 ganglioside. GM3 is the branching substrate for complex gangliosides. Once the gangliosides synthesized in the Golgi system are translocated to the cellular PM, GM3 is further utilized as a starting precursor molecule for sequential reaction in the a-series ganglioside biosynthesis. From the roles of GM3 in the initial synthetic substrate, it is suggested that GM3 is largely expressed during early embryonic development of the organism including the brain. After development progress likely in later steps, the GM3 synthesis is gradually decreased. However, GM3 synthase expression is constitutively maintained even during embryonic development for unknown reasons. For downstream synthesis, GM2 is converted from GM3 by the GM2 synthase (GalNAc transferase). GM1a ganglioside is further converted from GM2 via Gal linkage by Gal-transferase 2, known as GM1 synthase. Then, GM1a species is further modified to GD1a species via sialylation at the terminal sugar residue by a ST enzyme, termed GD1a synthase. Seemingly, the GT1a is also generated from the precursor GD1a by enzymatic catalysis of a ST GT1a synthase. Such enzyme species can also catalyze the enzymatic formation of the b–/c-series gangliosides because the starting precursor ganglioside is the GM3 [101]. Gangliosides are synthesized in the Golgi in orchestrated fashion of various glycosyltransferases, sequentially step by step adding sugar residues to the lipids. Of interests, gangliosides produced in nuclear regions modulate gene expression of certain targets. A marginal problem with neuronal ganglioside synthesis is that these are also produced in the ER-Golgi complex network or extracellularly internalized. How they are easily transported to the membrane lipid rafts? One answer is that lysosomes may provide such gangliosides, as mechanistically explained by the chaperone-mediated autophagy theory, proteasomal ubiquitinyl degradation, de novo synthesis, or recycling by autophagy. Collectively, from a precursor LacCer substrate, a series of complex GSLs such as ganglio series, globo, lacto, and asialo (GalNAcβ1-4Gal) GSLs are synthesized in cells. The metabolism and catabolism of gangliosides are described in Fig. 2.5.

Therefore, gangliosides are structurally complex molecules synthesized by the step-by-step attachment of monosugar residues or SA residues via each specific glycosyltransferase localized in the ER to Golgi apparatus. Gangliosides are diverse in molecules and structures of carbohydrates and ceramide structures. This property of gangliosides distinguishes the gangliosides produced by tumors from those produced by normal cells. Such structural heterogeneity indicates heterogeneous immunosuppressive activities of tumor gangliosides. Gangliosides overexpressed by cancer cells are frequently shed into the tumor-associated microenvironments (TAMs), influencing the immune escape behavior from immune surveillance [102].

Changes in ganglioside composition and distribution frequently appear in response to changes in cellular function and morphological alteration and function including tumorigenesis or other diseases status. For example, gangliosides induce apoptotic events. Representatively, ceramide-caused apoptosis is associated with gangliosides such as GM3 or GD3. From the changes in ganglioside synthesis upon cellular transformation, such gangliosides are used as potential diagnostic markers and therapeutic targets for the related cancers.

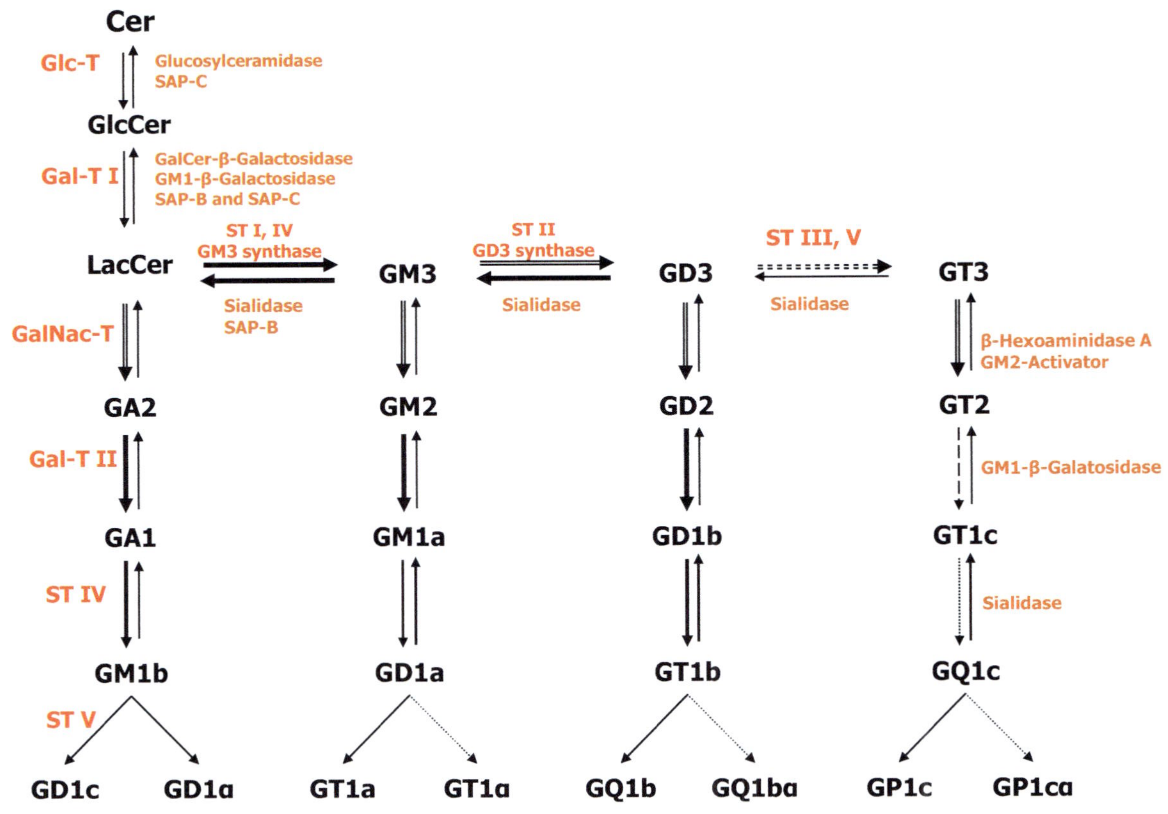

Fig. 2.5 The metabolism and catabolism of gangliosides are described. The sialyltransferases are abbreviated as the commonly used names. Ganglioside metabolic and catabolic pathways are modified from Huwiler et al. 2000, Biochim Biophys Acta 1485, 63–99 [103]

Chapter 3
The GSL-Dependent Signaling

GSLs recognize, bind, and interact with PM proteins including growth factor receptors (GFRs), tetraspanins (TSPs), receptor tyrosine kinases (RTKs), integrins, and nonreceptor cytoplasmic kinases including Src kinase family and G proteins in direct or non-direct mode. Such recognition and interaction contribute to the formation of glycosynaptic microdomains or lipid rafts to regulate GSLs-dependent cell behaviors such as attachment, adhesion, endocytosis, growth, and motility [17]. Gangliosides play key roles through glycan-targeted interactions with membrane receptor proteins (Fig. 3.1). In receptor-mediated biological events, they modulate cell-cell interaction frequently exploited by pathogenic agents such as viruses and bacterial toxins. They also modulate phosphorylating molecules via lateral binding to the membrane proteins of cells. For instance, they directly or indirectly inhibit such receptor-associated tyrosine kinases. In some cases, the interaction between specific gangliosides and cell surface receptor proteins needs posttranslational glycosylation of the PM glycoproteins. Such instances can explain the reason why GSLs-glycoproteins interactions are limited to specialized cases, but not to all the proteins. Gangliosides influence signaling of many GFRs which are currently well known. They include tyrosine kinase receptor (Trk) family, fibroblast GFR (FGFR), epidermal GFR (EGFR), basic FGFR (b-FGFR), platelet-derived GFR (PDGFR), nerve GFR (NGFR), insulin-like GFR (IGFR), insulin receptor, IGF-1 receptor, vascular endothelial GFR (VEGFR), glia cell-derived neurotrophic factor (GDNF), bradykinin 2 receptor (B2R), Fas (CD95) receptor, G protein-coupled receptor (GPCR), serotonin receptor, α1-adrenergic receptor (AR), amyloid-β (Aβ), estrogen receptor (ER), hepatocyte GFR (HGFR), transforming growth factor-β (TGF-β), α-amino-3-hydroxy-5-methylisoxazole-4-propionic acid (AMPA), and ionotropic glutamate receptor (AMPAR). For example, interactions of GM3-EGFR and GM3-specific RTK on cell surfaces are observed only in the glycosylation status of receptor proteins. Therefore, the GM3 seems to interact with N-glycosylated EGFR or glycosylated RTK to downregulate its RTK. However, specific ganglioside GM1 rather potentiates the neuritogenesis activity of neurotrophins, where gangliosides like GM1s enhance neurite genesis and formation in cultured primary neurons

C.-H. Kim, *Glycosphingolipids Signaling*,
https://doi.org/10.1007/978-981-15-5807-8_3

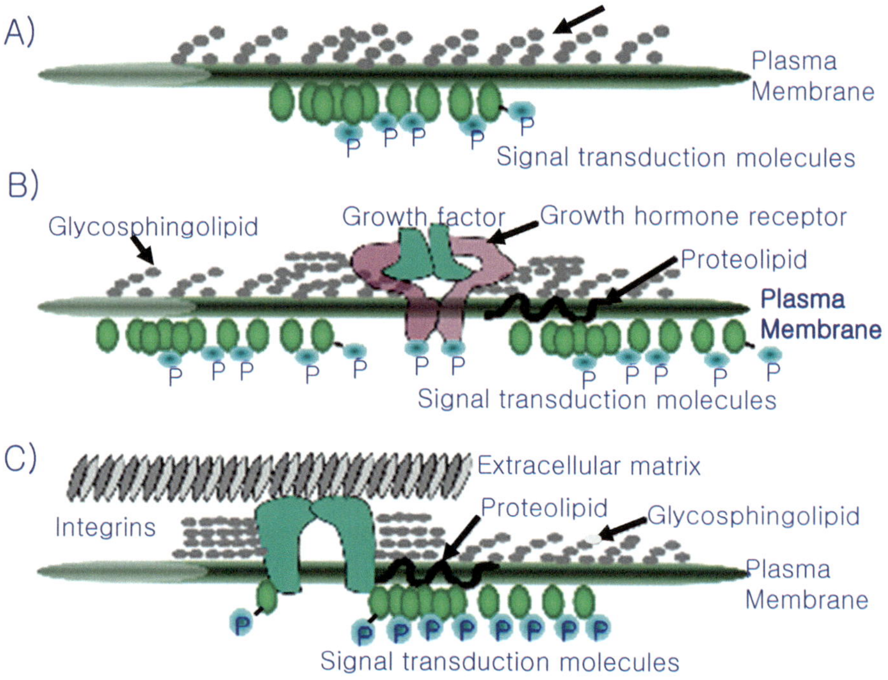

Fig. 3.1 The GSL-dependent signaling in cellular membrane integrins, growth factor receptors, receptor tyrosine kinases, and nonreceptor co-receptor signaling molecules in a direct or non-direct mode. (**a**) Normal state of cell membrane, (**b**) GSL-associated GRF-GRFR signaling status, (**c**) GSL-associated integrins and extracellular matrix signaling

and neuroblastoma cells [104]. GM1 also promotes the dopaminergic and GABAnergic neuron functions in cultured mouse mesencephalic cells. Treatment of the central nerve system (CNS) with GM1 mimics the NGF effect on the cholinergic neuronal protection upon cortical damage and the hippocampal regeneration effect of the CNS cells [105]. In the molecular mechanism, it has been suggested that GM1 protects the degenerative neurons by stimulating the Trk NGF receptor which can be applicable for other neurotrophic growth factors [106], promoting MAPKs and cAMP response element-binding (CREB) molecules in the axotomized nerve retinal region and promoting the dimerization of neurotrophic factor, RTK [107]. Of particular interest are some conflicting outcomes having opposite effects of some gangliosides on the same receptors depending on cell type or tissue type. For example, some gangliosides are overexpressed and shed from tumor cells. Such gangliosides bind to normal cells in the tumor microenvironment, altering tumor cells-host cells interactions for the survival environmental of the tumor cells. However, the present chapter focuses on the biological effects.

3.1 GM2/GM3-EGFR-RTK Interaction

GSLs regulate cellular signaling pathways by directly binding to acting molecules of the signal transduction ligands, receptors, and intracellular messengers [24]. One representative case is the EGFR-GM3 interaction [23, 108, 109]. The biological activity of EGFR is actively or negatively regulated by the GSLs distribution and composition in the membrane (Fig. 3.2). EGF when binds to EGFR activates the active homodimerization of the receptor protein from the inactive monomeric form. The dimeric form can activate the EGFR tyrosine kinase to trigger the EGFR autophosphorylation. The receptor activation contributes to downstream cascades including phosphorylation of transcription factors, nuclear translocation, gene expression of target genes, and changes in phenotypes including adhesion, migration, invasion, proliferation, and immune escape. Mutation of EGFR is frequently found from the cancers with tumor progression capacity [110]. Although receptor tyrosine kinases (RTKs) are activated by various growth factors, hormones, and cytokines, some specific GSLs promote or inhibit the RTKs functions. Autophosphorylation of EGFR is blocked by the exogenous and endogenous GM3 [71]. The molecular basis of the receptor interaction in vitro has been well explained using the liposome-based EGFR reconstitution. GM3 regulates the EGFR behavior by two independent ways. One is a PCI between the terminal NeuAc of the GM3 and a lysine (K642) to the EGFR, keeping the dormant EGFR state and inhibiting EGFR homodimerization. The second form is a CCI where the NeuAcα2,3-GM3 and the GlcNAc residues, which are terminally linked to *N*-glycans attached to EGFR, are recognized to bind. Seemingly, the EGFR function of phosphorylation and activation was blocked by exogenous and endogenous GM3 [111, 112]. The GM3 transcriptional regulation of the PTEN gene is also clarified [113] with detailed downstream antitumor function during GM3-EGFR binding in human colon cancer.

Unfortunately, the molecular protein-carbohydrate interaction (PCI) between GM3 and EGFR is not precisely explained, despite extensive studies. The inhibitory capacity of GM3 is dependent on the lipid moieties in GM3, since lysoGM3,

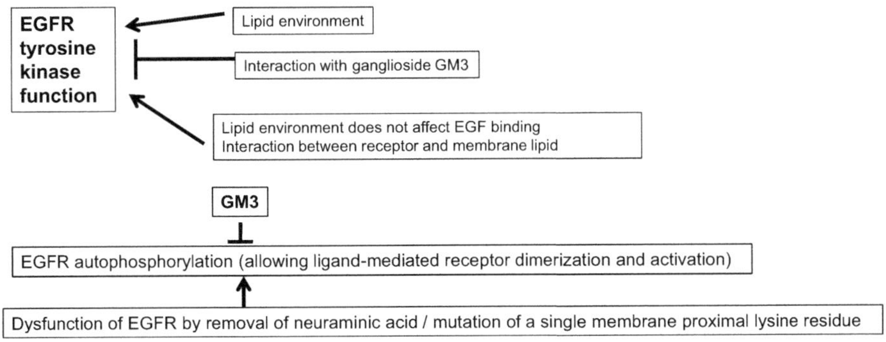

Fig. 3.2 The biological activity of EGFR tyrosine kinase is actively or negatively regulated by the membrane lipids [113, 123]

lysoGM3 dimers, and the mimic derivatives exhibited higher inhibitory activities than GM3 itself [109], but de-*N*-acetylated GM3 rather activated tyrosine kinase signaling of EGF receptor [114]. Only a fact that the GM3-mediated EGFR inhibition involves in the carbohydrate-carbohydrate interaction (CCI), where the gangliosides bind to the GlcNAc residues terminally attached to N-glycans on the EGFR, is suggested, as mentioned earlier. From the state of art, the additional scientific flow that membrane ganglioside organization for signaling affects the distribution, activation and phosphorylation of EGFR is updated. Then, the EGFR embedded in the membrane has been suggested to interact with the acidic lipid species [115–117]. For another action, gangliosides GM2 as well as GM3 are known to form complexes with membrane-specific RTKs, TSPs such as CD9, CD81, and CD82, and integrins [118, 119]. The formed ganglioside-membrane complexes suppress the receptor functions such as cell motility.

In contrast to the GM3-mediated EGFR downregulation, disialogangliosides of GD2 and GD3 are rather known to develop tumor-specific phenotypes. However, any relationship between such gangliosides and cancer development or aggressiveness is not well explained yet. For example, GD3 level is increased in the invasive cancer types such as ductal breast carcinoma cells. Interestingly, the α-N-acetylneuraminide α-2,8-sialyltransferase or GD3 synthase expression is correlated with the decreased survival rate in ER-negative breast tumor-positive patients [120]. GD2 and GD3 products levels with their glycosyltransferases of GD3 synthase and GD2/GM2 synthase also seem to keep a stemness and stem cell-like phenotypes of breast cancer stem cells (CSCs). This is because the GD3S keeps the sustained CSC phenotype in the type of malignant cancer phenotypes [121, 122]. GD3 was also known to be involved in EGFR to promote EGFR activation and signaling in breast CSCs and breast tumor cells, as this specificity of GD3 in EGFR is opposite to the GM3 behavior. Therefore, in a recent report GD3S silencing efficiently sensitized cytotoxic activity of the known EGFR inhibitor such as gefitinib in triple-negative and resistant breast MDA-MB468 cells. This clearly indicates that GD3S is an inducer of the gefitinib-resistant EGFR-positive breast cancer cells, potentiating a therapeutic possibility in drug-resistant breast cancers [120].

3.2 Gb4-EGFR Interaction

Apart from the functional roles of GM3, other gangliosides with the similar carbohydrate structure as GM3 include GM1, GD1a, and GT1b. They exhibit similar inhibiting activities against EGFR functions [114, 124]. Apart from the SA-containing ganglioside such GM3 or GM1, such inhibition has been counteracted by the Gb4 globoside, which is one of the major neutral GSLs [122]. As previously mentioned, a globo-series Gb4 GSL is another SSEA (Fig. 3.3). Gb3 is expressed in embryos of ESC and tumor tissues likely cancer SC without further information on the biological action and molecular mechanism. Gb4

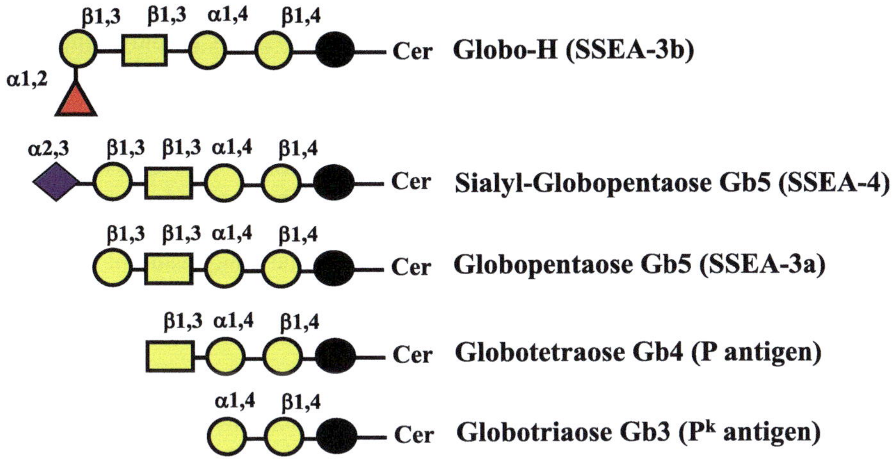

Fig. 3.3 Globosides of Gb3, Gb4, Gb5, and globo-H

produced in colon tumor HCT116 and breast tumor MCF7 cells activated ERK and EGFR potential in human RTKs via interaction between Gb4 and EGFR. Gb4 therefore contributes to cell development and tumorigenesis at the initial stages via previously known RTKs including EGFR, EphB2, FGFR3, HGFR, and Tie-2. Gb4 may bind directly to EGFR to trigger its activation and autophosphorylation. In addition, Gb4 further phosphorylates its downstream signaling molecules. Other neutral LacCer and Gb3 were negative for the EGFR effect [122], strengthening a specific glycan moiety structure and configuration to interact with receptors. However, unfortunately, the Gb4 binding site(s) on EGFR are not explained at the molecular level. Possibly, if CCI is applicable to the Gb4-EGFR interaction, such CCI possibility is deduced from the fact that Gb4 binds to both GalGb4 (Galβ3GalNAcβ3Galα4Galβ4GlcβCer) and nLc4Cer (Galβ4GlcNAcβ3Galβ4Glc βCer) [125]. For other cases, GM3 also negatively regulates the RTK activity in the FGFR [90, 126]; cross-talk interaction has also been suggested between integrins and FGFR which are modulated by GM3, as lack of GM3 promoted MAPK-mediated tyrosine phosphorylation of FGFR and Akt. Cell proliferation was enhanced by GM3 depletion.

3.3 GM1-Nerve Growth Factor (NGF) and GM1-Glial Cell-Derived Neurotrophic Factor (GDNF) Interaction in Neuronal Cells

In the brain, nerve growth factor (NGF) is known for its properties involved in survival and differentiation in certain types of sympathetic and sensory neurons. In the CNS of mammals, carbohydrate structures of gangliosides and distribution are tightly regulated. Among them, four monosialyl- to tri-sialyl gangliosides of GM1, GD1a, GD1b, and GT1b are dominantly produced [127]. Currently, only one example of the inherited diseases is known to be caused from a defected biosynthesis of gangliosides. For example, the defect of GM3 synthesis derived from active GM3 synthase deficiency is prevalent among the Old Order Amish populations [128] and is reported with clinically severe neurological syndrome. This GM3 synthase deficiency-derived human disease indicates an epileptic disease type in the Old Order Amish populations. Thus, simple ganglioside defect leads to the conclusion that many human diseases potentially develop from the ganglioside biosynthetic pathway dysfunction. Recently, another mutation has been found from GM2 synthase gene B4GALNT1 in an Italian-Canadian population as well as the Old Order Amish population [129].

In the beginning of studies, GSLs function and potential have been started from the neuronal tissue expression of GSLs and abnormal biosynthesis in the lysosomal storage diseases displayed by neuropathy. The gangliosides in nervous organ function in a mode of "trans" action together with the myelinating protein, which is called the myelin-associated glycoprotein (MAG) as a neuron-specific siglec type. MAG recognizes the NeuAcα2-3Galβ1-3GalNAc-terminal structures on gangliosides expressed in axon and the recognition is essential for axon–myelin maintenance and axonal regeneration. GM3 binds to GA2, gangliotriaosyl-Cer with the structure GalNAcβ1,4Galβ1,4Glcβ1,1-Cer and mediates adhesion. Globo-series Gg3 and ganglioside GM3 are known to specifically interact for the basic recognition for specific cellular communication between melanoma cells and lymphoma cells [130]. For example, MAG and oligodendrocytes' Siglec-4 bind to axonal GD1a and GT1b. The MAG-GD1a/GT1b binding evokes a negatively progressed transduction and prevents axonal outgrowth during neuronal tissue injury. This is the background that GSLs are for the firstly described case with NGF interaction for the cellular signaling. GM1 is known to stimulate the process of outgrowth of neurites of neuronal cells. However, CTx, a GM1-binding bacterial toxin, inhibits the outgrowth of neurites. In the neurotrophin receptors, GM1 ganglioside promoted neuronal growth, phenotypic change, and survival by regulating tyrosine kinase receptors for such neurotrophic pathways. Neurite outgrowth is initiated by a GM1-laminin interaction [131], indicating that GM1 functions as neurotrophic factor-like molecule. GM1 also stimulates the NGF-induced autophosphorylation of tropomyosin-related kinase A (TrkA) known as the NGF-activated RTK [132]. For example, it is reported that GM1 promotes neurite formation in neuroblastoma cells. The GM1 activates the signaling of TrkA-MAPK pathway. In addition, GM1 binds to regulatory proteins to modulate the biological events in the specific actions on ions transportation, neuronal cell differentiation, GPCRs modulation, immune responses, and neuronal protection. The neuroprotection event is exerted by interaction with neurotrophin receptors. Such outcomes designate the relevant etiopathogenesis of neurodegenerative diseases [133]. The Na^+/Ca^{2+} exchanger (NCX)/GM1 complex transfers nucleoplasm Ca^{2+} to the luminal area of the double membrane to protect the nucleus [134] through Ca^{2+} migration to the ER lumen region. The outcomes suggest that the oligosaccharide carbohydrates of GM1 is the acting site rather than Cer moiety.

Abnormal Trk-derived signaling has been related to multiple syndromes including AD, stroke, and amytrophic lateral sclerosis. Defects in the Tyr kinase domain of Trk in patients involve congenital pain insensitivity, an autosomal-recessive disorder. α-Trk autoantibodies are found in patients with subacute sensory neuropathy [135]. Therefore, Trk-derived signaling is important for the nervous systems. Trk-derived NGF signaling contributes to the Neurotropin efficacy. Therefore, neurotrophins are potentially applicable for the therapeutic target of neuronal diseases, but its low stability in serum, blood-brain barrier issue, and its pleiotropic binding to various receptors are problematic [136]. Thus, to overcome demerits of neurotrophins, NGF-mimicking small molecules with receptor specificity are candidates [137]. The candidate molecules function as Trk agonists to activate Trk signals. Neurotropin, unlike other Trk agonists, protects neurons only when target cells lack adequate tropism.

For GM1 synthesis, the known B4GALNT1 coding for the β1,4-N-acetylgalactosaminyltransferase 1 (GalNac-T) transfers GalNAc residue to GM3 or GD3 to yield GM2 or GD2 that further transferred to GM1 or GD1. Interestingly, GM1 and glycosylated Trk were bound together, protecting neuronal cells from neuronal injury and apoptosis [138]. Mechanistically, GM1 induces Trk dimerization upon NGF interaction. When the GM1-associated Trk induces NGF signaling, the molecule Trk is translocated into lipid rafted microdomains, where GM1 and other GSLs are present. The Trk-GM1 complex is formed dependently of NGF. The receptor-lipid interaction and Trk autophosphorylation by NGF are enhanced by neurotropin. In addition, receptor-lipid complex can be constructed independently of Trk autophosphorylation. Neurotropin potentiates rapid responses to NGF through the Trk-GM1 complex formation but the Trk-GM1 complex formation is not mechanistically explained yet. Rat pheochromocytoma PC12 cells treated with Neurotropin can form Trk-GM1 complexes rather than dimerization or Trk phosphorylation. During NGF induction, neurotropin stimulates the NGF-induced Trk autophosphorylation, strengthening the NGF signaling and the potential therapeutic importance of neurotropin for neurotrophin dysfunction [139].

GM1-glycosylated Trk recognition and interaction is a type of CCI, as known in GM3-EGFR binding. GM1 is required for the regeneration of neuronal cell in vivo and potentially beneficial for therapeutic drugs for Parkinson's disease and GM1 binds α-synuclein to uptake into cells, although the gene is frequently mutated in Parkinson's. Thus, Trk and α-synuclein are GSL binding proteins [140]. Potentially, asialo-GM1, GM2, and GM3 carbohydrates are negative for the neurite forming activity. Direct GM1 association with Trk promotes neurite outgrowth formation and neurofilament production stimulated by NGF in PC12 cells. Molecularly, GM1 stimulates NGF-dependent homodimerization of Trk in PC12 cells [139]. The importance of GM1 has been evidenced by the fact that GM1 depletion by GlcCer synthase inhibition suppresses the NGF-induced neurite outgrowth and the co-treatment with GM1 abolishes the effects in PC12 cells. Thus, the GM1 oligosaccharide directly activates the TrkA receptor and the downstream phosphorylation pathway, acting as a bridge affordable for the TrkA-nerve growth factor interactions [141]. In in vitro pheochromocytoma PC12 cells, the GM1 protects and has metabolic effects upon exposure to hydrogen peroxide, which is dependently expressed on activation signaling of the Trk RTK/Akt/ERK1/2 axis activation [142].

Ret tyrosine kinase receptor for GDNF has also been activated by GM1 in situ and in vivo on dopaminergic neurons. In addition, GM1 increases the binding of endogenous GDNF to GFRα1 co-receptor, which GM1 activates Ret, where Ret activation is derived by Tyr1062 and Tyr981 phosphorylation-based PI3-K/Akt, Erk, and Src signaling [143]. GDNF is a homologously conserved protein as a group of neurotrophic factors that are similar to the basic FGF group. The neurotrophic factors also include neurturin (NTN), persephin, and artemin as the same family [144]. The GDNF and its family factors utilize a GPI-anchored neurotrophic factor-binding subunit in PM lipid rafts and they activate the C-ret/c-Src Tyr kinases. Trophic failure of catecholaminergic neurons causes parkinsonism or PD. PD is also linked to signaling defects caused by the catecholaminergic growth factor, GDNF. In addition, PD has been hypothesized to be linked with the synthesis deficiency of a series ganglioside GM1 [145]. The decreased synthesis level of GM1 is suggested to involve in parkinsonian degeneration and therefore, GM1 supplementation ameliorates the status of PD outcome. Regularly expressed striatal GDNF normalizes the α-synuclein levels by eliminating the aggregated forms of alpha-synuclein in nigral neurons. Consequently, the GDNF expression restores motor neuronal action in the neurons. Interestingly, GM1-GDNF interaction diminishes the status of α-synuclein pathology in PD, as demonstrated in the results that fibrils are formed in the absence of GM1 because the interacting motifs observed in α-synuclein is a ganglioside-binding protein [144]. The binding motif (34-KEGVLYVGSKTK-45 sequence) is also found in prion protein and β-amyloid protein [146]. The binding region of α-synuclein is a helical structure and the binding site-disrupting mutations in the α-synuclein damage the GM1 binding property. The binding domain contains a tyrosine residue (Y39) to be phosphorylated by c-abl. Additionally, the two non-equivalent cholesterol-binding domains (VLVYVGSK) are also found in the primary sequence of α-synuclein. One domain is located on the GM1-binding motif and another cholesterol-binding site is the motif (67-GGAVVTGVTAVA-78).

3.4 GT1b and GD1b-B2R Interaction in Neuronal Cells

Gangliosides stimulate differentiation and maturation of neuronal cells for the normal function and survival of brain. Figure 3.4 illustrates the total structures of ganglioside a/b/c-series and sialyltransferases. For the mechanistic internalization of exogenous gangliosides into target cells, it has been suggested that gangliosides such as GT1b form micro-micelles or vesicles in aqueous solutions and this property potentially enables to adhere to the cell surfaces by the micelles, increasing in the diffusion barrier levels [147]. GT1b stimulates calmodulin-dependent protein kinase

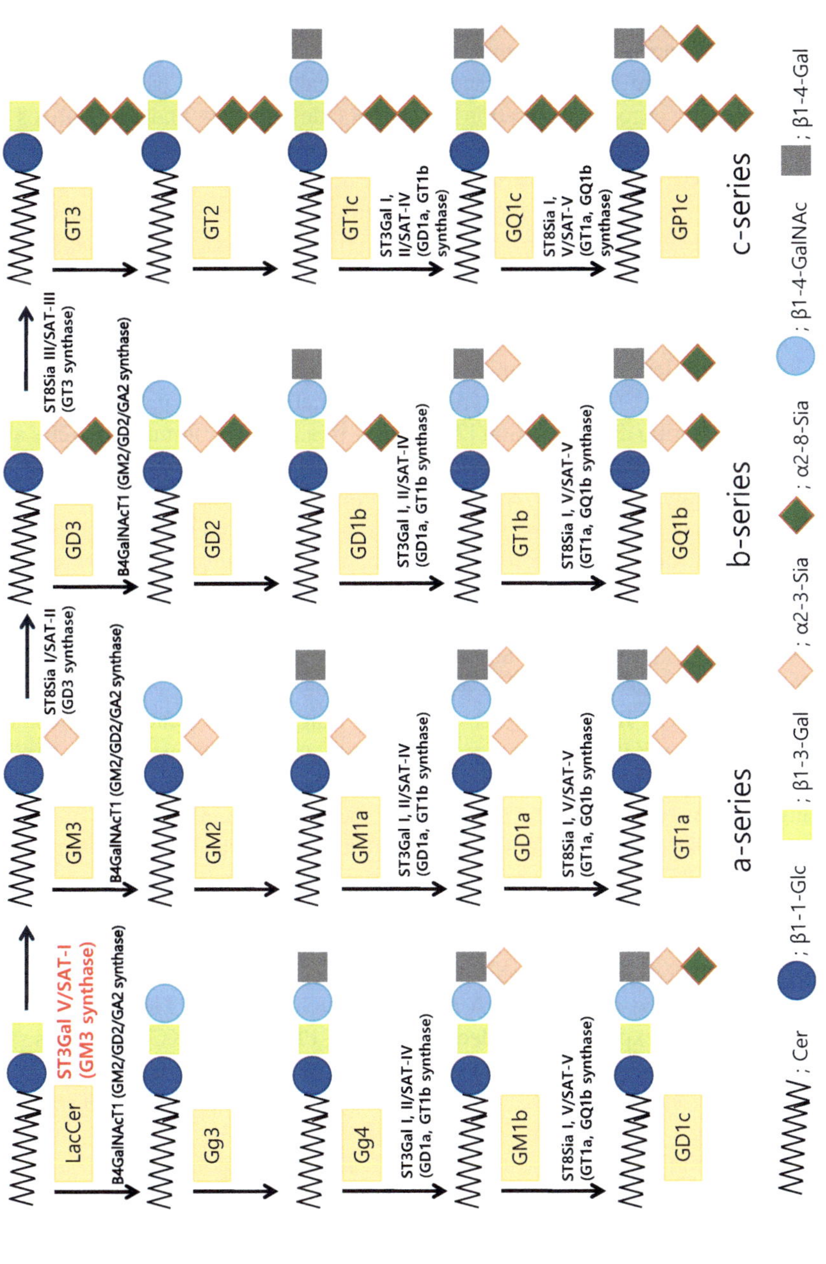

Fig. 3.4 Structures of ganglioside a, b, and c-series and sialyltransferases

II (CaMKII), which is distributed in synapses with a capacity of cdc42-mediated cell cycle activation. GT1b activates B2R downstream signaling pathway, as has been confirmed in yeast cells that recombinantly express mammalian B2R gene [148]. The B2R is one of GPCRs located on PM and it is coupled to accessary molecules of Gq and Gi. Among the two G proteins, G_q activates phospholipase C (PLC) to increase the level of intracellular free calcium ions, while G_i inhibits adenylate cyclase. Agonist activation of GPCRs desensitizes and internalizes to render them unresponsive to agonist activation. B2R induces filopodia formation and differentiation of axon- and dendrite-like neurites. The b-series ganglioside GT1b stimulates B2R during neuronal differentiation in mice, affecting the Ca^{2+} concentration [149]. Neuronal cells react to GD1b, βDGalp1,3βDGalNAc1,4 [αNeu5Ac2,8αNeu5Ac2,3]βDGalp1-4βDGlcp1,1Cer and GT1b, αNeu5Ac2, 3βDGalp1,3βDGalNAc1,4,[αNeu5Ac2,8αNeu5Ac2,3]βDGalp1,4βDGlcp1,1Cer, largely present in the synapse-forming region of the mammal brains. GT1b and GD1b increase the level of Ca^{2+} ion release from intracellular Ca^{2+} ion storage small organelle, activate Ca^{2+}/CaMKII enzyme, and stimulate cdc42 for cytoskeletal actin structure and dendritic differentiation [2]. The GT1b and GD1b stimulated B2Rs as a signal transducing mediator to transduce the glycan-mediated signaling pathway. A known B2 antagonist, Hoe140, specifically inhibits GT1b and GD1b-stimulated CaMKII functional activation, axon reorganization of actin and actin axon development as well as dendrite-genesis of primary hippocampal neurons. In yeast reporter assay, GT1b, GD1b, and GD3 stimulated B2Rs, indicating that B2R-mediated ganglioside signal transduction pathway induces neuronal differentiation and maturation. CaMKII is a family of the Ser/Thr protein groups and its 28 isoforms, which are originated from the four distinct genes including the α, β, γ, and δ genes, are identified [150]. CaMKII recognizes the Ca^{2+} signaling-regulating molecules. CaMKIIγ and CaMKIIδ are previously reported to express in various tissues [151]. Ca^{2+} cellular influx is crucial for functional stimulation of downstream signaling proteins, which include CaMKIIγ in meiotic resumption raised by sperm and in embryonic development. B2R stimulates the PLC and PKC downstream signaling. In addition, B2R further activates the independent downstream cAMP and PKA signaling pathways. However, cAMP inhibits cell proliferation. Activation of the independent cAMP and PKA pathway suppresses B2R signaling and reduces bradykinin-dependent inositol triphosphate and Ca^{2+} mobilization [152]. The B2R expression is inhibited if cAMP/PKA pathway is treated with GT1b. Therefore, if the level of B2R expression is decreased, the PLC and PKC pathways are downregulated. The question how gangliosides activate the B2Rs has been explained by the results that gangliosides are not the possible stimulator of intracellular Ca^{2+} release from B2R-positive CHO-K1 cells [153]. However, multiple treatments of gangliosides desensitized the B2R responses, inducing efficient internalization of B2Rs. Gangliosides likely induce desensitization of B2R signaling and intracellular endocytosis via the B2R phosphorylation.

3.5 GQ1b-N-Methyl-D-Aspartate (NMDA) Receptor Interaction for Long-Term Potentiation (LTP) and Brain-Derived Neurotrophic Factor (BDNF) Synthesis in Neuronal Cells

Gangliosides are involved in neurogenesis, proliferation, synaptogenesis, and synaptic transmission. Among the gangliosides, GQ1b is neurogenic on human neuroblastoma cells and stimulates neurogenesis through the receptor protein phosphorylation of neuronal cells [154]. ATP-induced LTP upon GQ1b treatment induces LTP at hippocampal CA1 synapses. Monosialyl GM1 and tetrasialyl GQ1b gangliosides exhibit the activities of ATP-elicited long-term potentiation (LTP) in neurons in the brain CA1 region of guinea pig hippocampal slices (Fig. 3.5). LTP is previously defined as a persistently enhanced synaptic behavior, which occurs by a high-frequency electric stimulation (HFS) of afferents during memory, education, and learning [155]. In hippocampal CA1 neurons, the LTP in memory, learning, and education is activated through the N-methyl-D-aspartate (NMDA) receptor with the induction of downstream second messengers [156]. NMDA receptors are classified to the glutamate-gated cation channels involved in education, learning, and memory potentials. NMDA receptor-linked Ca^{2+} influx can trigger a variety of downstream signaling (Fig. 3.6). The extracellular signal-regulated kinase (ERK) and the ERK downstream have been regulated through phosphorylation at the amino acid Ser-133 position of cAMP responsive element binding protein (CREB) [157]. CREB cis-upregulates the target gene expressions including cAMP response elements (CRE). They regulate neuronal survival, proliferation, and differentiation as well as LTP induction and maintenance with memory and learning capacities. For example, one of the target genes, which CREB upregulates synaptic plasticity, is the BDNF, which belongs to the TGF-β superfamily, keeping the catecholaminergic neurons [158]. GM1 interacts with the GDNF receptor complex, which is associated with the Tyr kinase Ret and GPI-anchored GFRα co-receptor. GFRα recruits the Ret

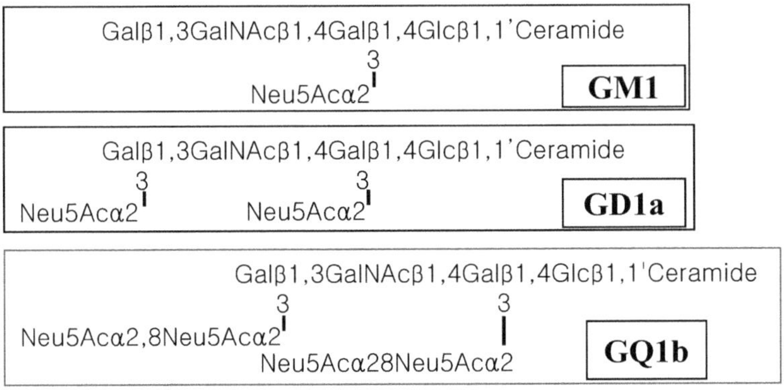

Fig. 3.5 Structures of GQ1b, GD1a, and GM1

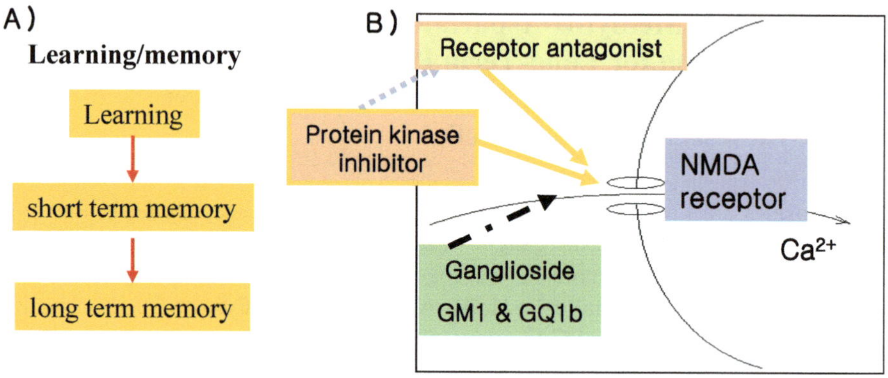

Fig. 3.6 (**a**) The major two steps of experience-acquisition for education, learning, and memory in human brain. The first short-term memory step but unstable due to temporal neuronal action and the second long-term memory step stabilized for neural adaptation to experience-acquired gene expression. (**b**) In the brain CNS, gangliosides potentiate neuronal activities including neurogenesis, synaptogenesis, and synaptic transmission. Gangliosides take part in LTP through the NMDA receptor/Ca^{2+} ion channel functional activation. LTP is induced and activated mainly by GQ1b as well as minorly by GM1 and GD1a in CA1 neurons of the hippocampus. This has been evidenced by the results that a NMDA receptor antagonist-blocked LTP induction is reversed by GQ1b treatment. Then, Ca^{2+} ions are released from the cells

to microdomain lipid rafts in membrane upon GDNF recognition [159]. Therefore, GM1-synthesizing enzyme, B4galnt1-deficient mouse, is impaired in the Ret-GFRα interaction [160].

BDNF belongs to a family of the neurotrophins. The neurotrophins known are NGF, neurotrophin-3, and neurotrophin-4/5. BDNF also involves in impaired memory diseases of Alzheimer's disease (AD) and PD. PD is improved by GM1 therapy to partially restore striatal dopamine lost and nigrostriatal neuronal recovery in PD model animal [161, 162]. GM1-mimic compound named LIGA20 has been designed and its efficacy has been reported to be superior to GM1 in the mice model [163]. PD patients who received subcutaneous daily GM1 are improved for PD evaluating motor scores in the phase II trial [164]. The endogenous GM1 is crucial as demonstrated by B4galnt1-KO mice [165] and the heterozygotes (B4galnt1+/−) [166]. GM1 deficiency raises the impaired GDNF signaling and retards phosphorylated Ret in the mouse and PD patients. GFRα receptor-Ret complex formation elicited by GDNF is also impaired depending on the GM1 deficiency [167]. Thus, if membrane-permeable agents are developed as analog types of GM1 ganglioside, they may be used as candidates for PD because GM1 replacement therapy would restore long-term, constitutive GDNF signaling with the GDNF receptor. GM1 efficacy in PD is related to α-synuclein [161] through acetylation of the N-terminal region of α-synuclein, which enhances GM1-interaction capacity [168]. The glycolipid-binding domain present in α-synuclein consists of a loop centered Tyr (Y39) residue and can bind to GM1 and GM3 [169].

BDNF-deficient mice exhibit dysfunction in spatial learning and memory performance [146]. Seemingly, LTP activation needs glutamate-promoted Ca^{2+} influx via NMDA receptor channels by phospholipase C (PLC) activation of the metabotropic glutamate receptor (mGluR). Consequent inositol triphosphate (IP3) and diacylglycerol (DAG) are generated by the PLC enzyme from phosphatidylinositol 4,5-bisphosphate (PtdIns4,5P2) known as PIP2 (Fig. 3.5). In addition, Ca^{2+} is also released from ER stores of the cells. LTP is maintained by the certain second messengers or Ca^{2+}-induced protein kinases. The second messengers currently known are arachidonic acid and cAMP. The induced protein kinases phosphorylate downstream signaling proteins such as AMPA and NMDA receptors. In cultured cortical neurons, the biosynthesis of gangliosides is inhibited by a Cer-mimic, d-threo-1-phenyl-2-decanoylamino-3-morpholino-1-propanol (D-PDMP) but is activated by its l-enantiomer (L-PDMP) [170]. L-PDMP enhances the synapse formation in primary cortical neuronal cells. For adaption, L-PDMP administration restores the in vivo spatial memory potentials of ischemic animals such as rat. Therefore, the relationship between the memory enhancement and generation of L-PDMP-induced b-series gangliosides is strengthened in neurons. The ganglioside b-series include GD3, GD1b, GD2, GT1b, and GQ1b, generated by neuronal cells. The enhanced potentiation of synapses formation and synaptic activity is therefore controlled by ganglioside biosynthesis such as oliosialyl GQ1b [155]. Gangliosides-depleted cells can also be restored to acquire the synaptic activity when GQ1b is added due to GQ1b-mediated synapse formation and activity. GQ1b has been known to enhance ATP-induced LTP and cognitive functions in CA1 neurons [171]. For the molecular mechanism of the GQ1b-involved LTP, GQ1b directly involves in the NMDA receptor/Ca^{2+} channels regulation. The cationic Ca^{2+} is bound by anionic GQ1b. Among the negatively charged gangliosides, GM1 consists of only one negatively charged SA residues. However, GQ1b consists of four SA residues, which recognize positively charged Ca^{2+}. Interestingly, LTP can also be stimulated by monosialyl GM1. For example, GM1 activates LTP in Ca^{2+} – enriched medium. LTP induction is enhanced by GQ1b but not with GM1 during HFS [171]. Antagonists of NMDA receptor diminishes the LTP induction effect, while the regressed LTP is reversed by GQ1b treatment. Therefore, GQ1b keeps the normalized synaptic activity through activity-dependent LTP maintenance with the downstreamed NMDA receptors/Ca^{2+} channels regulation.

On the other hand, GQ1b enhances BDNF biosynthesis in the genetically GQ1b-deficient SH-SY5Y cells and primary cortical neurons of rats. From the fact that an NMDA receptor antagonist D-AP5 blocks the GQ1b-enhanced BDNF expression, GQ1b is considered to regulate the NMDA receptor signaling for the BDNF expression. In vivo intracerebroventricular injection of GQ1b enhances the BDNF production in the hippocampus and prefrontal cortex in the experimental rats. The intracerebroventricular injection of exogenous GQ1b increases in prefrontal and hippocampal BDNF production [172]. Interestingly, GT1b or GD1b treatment is effective for the BDNF expression, whereas GQ1b upregulates the BDNF production. GQ1b directly influences the BDNF production in the established SH-SY5Y cell and primary cortical neurons of rats. Moreover, GQ1b upregulates BDNF expression in protein and gene levels. Furthermore, GQ1b activates the NMDA receptor signaling to regulate the BDNF expression in cultured cortical cells of rat primary neurons. The prefrontal and hippocampal expression levels of BDNF are in vivo increased by ICV injection of GQ1b in naïve rats. GQ1b restores the BDNF level decreased by the D-PDMP, a ganglioside biosynthesis inhibitor, in primary cortical neurons of rats. However, the NMDA receptor antagonist inhibits the GQ1b-mediated BDNF expression. Therefore, GQ1b regulates the NMDA receptor to activate the downstream signaling toward the BDNF expression.

GQ1b shows neurotrophic activity in vitro and in vivo with diverse neuronal functions. GT1b, GD1, and Cer are not inducible for BDNF protein synthesis. Even though GM1 slightly upregulates BDNF expression, GQ1b was prominent for the BDNF expression. GQ1b upregulates BDNF expression capable for synaptic plasticity. GQ1b enhances learning and memory in animal behavior assessed upon intracerebroventricular injection of GQ1b by the Y-maze and MWM tests [173]. GQ1b-administered rats exhibit enhanced improvement of performance, not affecting basal locomotor activity. For the improvement of GQ1b functions to AD progression, treatment with Aβ causes AD. However, GQ1b treatment restores Aβ1–42-caused neuronal cell death through the enhanced BDNF expression in cortical neurons in a primary culture. GQ1b injection to the hippocampal region improves cognitive impairment and deficits in the triple-transgenic AD mice. The GQ1b-infusion to the triple-transgenic AD mice increased BDNF expression levels. In addition, GQ1b hippocampal treatment in the triple-transgenic AD mice largely decreased the levels of Aβ plaque deposition and tau phosphorylation, which are previously known to be linked to a decrease in amyloid precursor protein level and an increase in phospho-GSK3β level, respectively [174]. Changes in ganglioside composition are observed in the AD patient brain. The decreased levels of certain types of GM1, GT1b, GD1b, and GD1a gangliosides are observed; however, GD3, GM2, and GM3 levels are increased in AD development [175–178].

For the AD development, GM1 recognizes Aβ functioning as an endogenous Aβ assembly factor [179, 180]. Aβ1–40 co-treatment with GM1 is cytotoxic for mouse neuroepithelial cells [181]. A mutant APP-expressing transgenic mouse, whose GM2 synthase gene was genetically disrupted and GM3 accumulates but GM1 lacks, showed a dramatically deposited Aβ level in the cardiovascular cells and tissues. GM3 induces the aggregated Aβ deposition [182]. In the brain tissues of AD patients, GD1a is highly expressed with phenotypes of dystrophic neurites and senile plaques, contributing to Aβ plaque formation [183]. Interestingly, among gangliosides, GQ1b level is uniquely decreased in human AD patients and mouse AD models [175, 184].

Conclusively, GQ1b enhances LTP, cognitive improvement, AD protection, and BDNF expression. For example, GQ1b-increased BDNF expression is beneficial for the AD models with cognitive impairment diseases, tau and Aβ physiopathology in AD patients and models. GQ1b plays a key role in synaptic plasticity and cognition because it exerts synaptogenic and neuritogenic activities in GQ1b-silenced neuroblastoma cells or primary cortical neurons. Because the GQ1b is essential for synaptic activity and LTP is the fundamental basis of memory, education, and

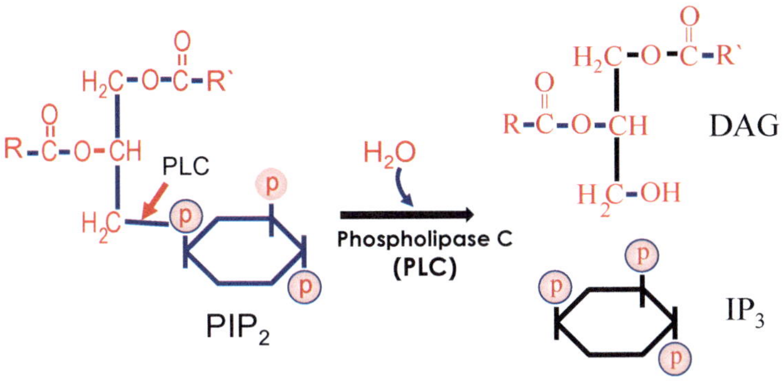

Fig. 3.7 Phospholipase C (PLC) generates inositol triphosphate (IP3) and diacylglycerol (DAG) from phosphatidylinositol 4,5-bisphosphate (PtdIns4,5P2) known as PIP2

learning at the cell level, the enhanced de novo biosynthesis of largely sialylated gangliosides is regarded as the potentially available treatment of various deficits in education, memory, and learning during aging (Fig. 3.7).

3.6 GM3/GM2/GD1a-HGFR Interaction

In the HGFR [185–190], GM3 or GD3 has been reported to stimulate the behavioral HGFR/cMet-positive hepatoma cells. However, in the certain condition, GM3 can exhibit its opposing effects by many different GFRs on the cellular migration in hepatocellular carcinoma Hepa1–6 cells of the mouse, depending on EGFR and HGFR/cMet. GM3 exhibited the inhibition of EGF-induced motility and the activation of HGF-induced Hepa1–6 cell motility, which is mediated by Akt and phosphatidylinositol 3-kinase (PI3K) signaling. GM3 acts on cells in a tumor cells-dependent way [188, 191]. GM3 influences cMet downstream signaling and HGF-stimulated cell migration capacity. GM3 further regulates cell motility in mouse hepatic tumor cells such as HcaA2, Hca16A3, and Hepa1–6 [189]. Lack of GM3 expression blocks the HGF-induced Met and PI3K/Akt phosphorylation during signaling pathways. With the exogenously treated GM3, the HGF-promoted phosphorylation levels of PI3K/Akt and cMet signaling activities are recovered. In addition, HGF-induced cell migration level and motility were blocked upon reduction of GM3 level. Therefore, for the activity of GM3 in HGF-induced motility of hepatoma cells, the enhanced phosphorylation of cMet and PI3K/Akt is the key regulator for stimulation of the migration and motility signaling in the cells.

In the opposite effect, c-Met (HGF receptor) is activated by GD3 synthase activity to activate the proliferation of aggressive breast tumor MDA-MB-231 cells [188, 189]. GD3 synthase modulates epithelial-mesenchymal transition (EMT) and **CSC** character. GD3 synthase regulates stem cell function via c-Met and the EMT signaling pathway's transcription factor FOXC2 regulated the GD3 expression. EMT/CSC-specific triple negative human breast cancers upregulate GD3 synthase gene expression with poor survival. GD2 and GD3 are regarded as a novel breast CSC marker and key driver for glioblastoma stem cells, respectively [187, 189]. GD3 synthase-c-Met signaling axis is considered to act as the future target for therapy of the malignant breast cancers. In addition, because GD2 is converted from GD3, GD3 serves as a potential target for tumorigenesis and metastasis. For mechanistic side, however, HGF/Met signaling was not directly activated by GD3 expression. Only when HGF was added, tumor phenotypes could be found in GD3-positive cells. Thus, it seems that GD3 plays a key role synergistically for the signals.

In addition, it was evidenced that GD1a suppresses the HGF-exerted motility of the highly metastatic FBJ-LL and HepG2 cells by inhibition of c-Met-induced Tyr phosphorylation. Thus, GD1a negatively regulates c-Met activity in tumor cells [192]. GD1a diminishes the ligand-recognizing motif on c-Met domain or alters the c-Met structure. Because GD1a ganglioside is relatively low in its molecular weight (MW 1836), its carbohydrate chain exists in the extracellular region; while c-Met protein is a relatively high molecular weight protein, consisting of two 50 kDa α- and 145 kDa β-chain subunits, it is difficult for GD1a to directly obstruct the ligand-binding motif of c-Met. From the circumstance, it is concluded that GD1a blocks c-Met dimerization via direct binding to c-Met [192]. Moreover, GM2-CD82 tetraspanin complex, but not GM2, GM3, or CD82, suppresses HGF-stimulated tyrosine kinase of cMet in the glycosynaptic microdomain of HCV29 cells and epithelium [118, 119]. Such inhibition on cell motility was found by blocking of cMet activation induced by c-Met. Cell adhesion is also blocked by the GM2-GM3-CD82 complex via obstructing integrin-mediated signaling pathway.

3.7 GM1/GM2-PDGFR Interaction

PDGF is well known to respond to cellular developmental events via certain signaling pathways such as MAPKs, PI3Ks, and PLC. In downstream signaling, PLC recognizes the phosphorylated form of PDGFR to activate it and generate diacylglycerol (DAG) and IP3 which cause for Ca^{2+} release from ER. PDGFRs are also subject to be regulated by cell surface or exogenous gangliosides, which are expressed by selfish or neighboring cells. In ganglioside regulation of PDGFR [193–197], PDGF-B in astrocytes increases the GD2 and GD3 expression levels. GD3 increases in cell growth and metastasis potentials with phosphorylation of Akt and Yes kinases. Binding between PDGF and PDGFR provokes its RTK to autophosphorylate its own tyrosine residues. The p85 subunit of **PI3K** as regulatory part targets phosphor-Tyr residues on PDGFRβ and the assembled p110 subunit of PI3K as the catalytic part catalyzes the formation of lipids to do intracellular signaling. PDGFB homodimer binds to PDGFR dimers of PDGFRαα, PDGFRαβ, and PDGFRββ, to contribute to RAS/MAPK/PI3K/Akt/JAK/STAT axis signaling pathways [195]. PDGF signaling pathways upregulate the expression of PDGFRα to promote oncogenic phenotypes. PDGFRα has been known as a GD3-interaction molecule in astrocytes, as GD3 colocalizes in GEM/rafts with PDGFRα. GD3 recruits PDGFRα to form GEM/rafts cluster. In human glioma, PDGFRα expression on the cell surface activates RAS/MAPK/PI3K/Akt axis signalings [196]. Thus, blocking the ligand binding to PDGFRα displays the blocked downstream signals and therapeutic avenue. PDGFRα has been associated with ganglioside GD3. GD3-producing astrocytes exhibit PDGFRα expression compared to the GD3-deficient astrocytes. Yes kinase, a Src kinase family, is also associated with PDGFRα. PDGFRα and GD3 were directly associated, forming GD3 complex with PDGFRα and Yes. The complex of GD3, PDGFRα, and activated Yes of tumor cells and glioma tissues explains that PDGFB-stimulated GD3 activates PDGF signaling pathways in glycolipid-enriched microdomain and lipid rafts with malignant phenotypes in gliomas.

In a similar way, the ganglioside GM1 in Swiss 3T3 fibroblasts inhibits both PDGF-triggered Tyr phosphorylation of PDGFRβ and endocytosis mediated by PDGFRβ receptor into the cells. However, GM1 alone cannot exhibit any efficiency on receptor phosphorylation, neurite-genesis, or endocytosis when PDGF has been treated in the PDGFRβ-transfected PC12 cells. The different action of GM1 to PDGFRβ in the two different cells of 3T3 cells and PC12 cells implies that GM1 acts in a cell context-dependent manner. The GM1 inhibits the PDGFRβ function in Swiss 3T3 fibroblasts by interaction with the extracellularly protruded transmembrane domains of PDGFRβ [198]. The question whether GM1 affects interactions between PDGFRβ and adaptor proteins to display PDGFR-mediated signaling is important at the initial studies. If so, PDGF-induced assembling of PDGFRβ with ras GTPase-activating protein, p85, and phospholipases are anticipated. Interestingly, GM1 reduced the interactions with reduced Tyr phosphorylation of PDGFR in cultured human glioma cell. GM1 inhibits PDGF-induced signaling proteins with PDGFRβ with suppressed phosphorylation of receptor Tyr residues bindable to signaling protein SH2 domains [199]. Gangliosides GM1, GM2, and GM3 inhibited PDGF-stimulated vascular smooth muscle cell (VSMC) growth and the PDGF-BB-induced signaling pathway in aortic VSMC of rats. GM2 and GM1 diminish the autophosphorylation of PDGF-BB-mediated RTK with the downstream signals including the inositol-1,4,5-trisphosphate (InsP3), PLC-γ1, free Ca^{2+} in the cytoplasmic region, c-fos, and cell proliferation. However, GM3 does not affect the PDGF-BB-mediated RTK autophosphorylation and PLC-γ1 activation. GM1 and GM2 efficiently suppress the specific binding of PDGF-BB, while GM3 is not effective for the PDGF-BB binding. Thus, GM1 and GM2 recognize the PDGF-BB or its receptor to prevent its binding.

3.8 GM3-VEGFR Interaction

Three subtypes of VEGFR-1, 2, and 3 are known. Membrane-bound VEGFR or soluble VEGFR is produced, depending on alternatively spliced variation. There are two known RTKs for VEGF, which are FLK1/KDR (VEGFR2) and FLT1 (VEGFR1). VEGF and VEGFRs of FLT1 and FLK1/KDR are crucial angiogenic factors. For the VEGFR [200, 201], several membrane ligands are known to regulate the EGFR and VEGFR signaling. GM3 is known to inhibit VEGFR-2 function through VEGFR dimerization [201] on umbilical vascular endothelial cells. GM3 recognizes the extracellular domain (ExD) of VEGFR-2 as a type of the CPI [202]. GM3 blocks VEGF-activated VEGFR-2 function in vascular endothelial cells. GM3 blocks neovascularization events through a PCI of GM3-extracellular domain of VEGFR-2. In contrast, GD1a rather induces VEGFR-2 signaling [203]. The proangiogenic GD1a is reversely reduced by GM3 [204]. In addition, FLT1 is expressed in a soluble isoform named sFLT1 (100 kDa) or full-length isoform (Flt1 170 kDa). sFLT1 recognizes the GM3 present in microdomain lipid rafts in the cellular PM surfaces of glomerular podocytes. This sFLT1-GM3 interaction consequently activates adhesion and actin filament reorganization [205]. GM3 downregulates several receptors including EGF, IGF-1, PDGF, b-FGF, VEGF, and cell adhesion molecules such as the integrins [206]. A group of gangliosides of GM1/GM2/GD1a/GD1b/GD3/GT1b stimulates the angiogenic inducing activities, as the opposing aspect of GM3 [207].

Tumor-associated macrophages (TAMs) involve in the immune suppression, angiogenesis, metastasis, and growth of tumor cells [208, 209]. Macrophages are differentiated from circulating monocytes and recruited by chemotactic factors at tumor sites [210]. Tumor-shed gangliosides induce angiogenesis, whereas GM3 exerts anti-angiogenic or angiostatic effects [201]. The tumor angiogenesis level is decreased in ganglioside-depleted tumor cells-injected mice, which bear double-KO of SA-transferase 9 and Gal-T1 [211]. Tumor-infiltrating macrophages express GM1b, asialo-GM1, and GD1α as o-series gangliosides when compared to the peripheral macrophages [212]. GM3 also suppresses the VEGF-activated synthesis of vascular cell adhesion molecule (VCAM)-1 and intracellular CAM-1 (ICAM-1) in HUVECs [213]. GD1a has been known to enhance the growth and migration of VEGF-induced HUVECs [214]. Certain and complete ganglioside deficient-tumor cells with the double deletion of the GM3 synthase-encoding Siat9 gene and GM2 synthase-encoding Galgt1 gene result in the reduced growth phenotypes, compared to the normal tumor cells [211]. GM3 inhibits growth factor recognition with regular stromal cells to respond [207, 212, 215].

3.9 GM1-FGFR Interaction

Gangliosides as modulators of protein phosphorylation regulate cell growth. In neurogenesis, recovery from nerve injury and neurodegeneration from diseases such as Parkinson's disease is not yet clearly answered. The neural CAM (NCAM) is specifically involved in the promotion of neurite outgrowth by receptor signaling of the FGF. NCAM translocation to membrane lipid rafts and neurite outgrowth stimulation of hippocampal neurons are performed by FGF receptor (FGFR)-FGF2 interaction [216]. FGF2 is a heparin-binding protein with a MW of 18,000 and the activities to induce endothelial cell proliferation, migration, and proteases production and neovascularization [217]. FGF2 activates FGFR-1 to −4 receptors, where FGF2 recognizes endothelial cells by receptor binding of tyrosine-kinase receptor FGFR, heparan sulfate proteoglycans (HSPGs), and αvβ3 integrins [218]. Free gangliosides, but not membrane gangliosides, can display opposite effects, regulating the growth factor and cytokine effects. For example, free gangliosides suppress PDGF, insulin, NGF, and IGF-stimulated neurite outgrowth and proliferation [194, 219]. In contrast, cellular PM gangliosides stimulate EGF, FGF2, and PDGF-promoted fibroblast growth [220]. FGF2 also induces axonal growth via glucosylceramide synthesis [221]. To exert the activity, the gangliosides directly bind to the growth factors and the example is the interferon and ganglioside interaction [222]. Cellular PM GM1 directly binds to FGF2. In addition, GM1 is also a co-receptor responsible for the growth of endothelial cells [223]. The FGF2 interacts with gangliosides even in in vitro cell cultures [224]. FGF2-bound gangliosides block the FGF2-FGFR interaction, consequently preventing the FGF2-induced endothelial cell growth. However, cellular PM gangliosides stimulate FGF2-induced fibroblast growth.

In a recent report [225], ganglioside GM1 stimulates growth and differentiation of the neural stem cell (NSC). GM1, but not a neural growth factor, significantly elevated NSC growth. GM1 treatment increased the cell growth, but not NGF treatment with the NSC marker nestin expression and glial fibrillary acidic protein (GFAP), which is a glial cell biomarker, and neuron-specific enolase as the neuronal markers. Therefore, it has been suggested that GM1 induces growth and differentiation of NSC in cultures. In addition, bFGF synergistically with GM1 stimulates the transformation of BM stromal cells, neuronal differentiation, and astrocyte differentiation. Such promoting effects have independently been observed by bFGF, GM1, or bFGF plus GM1 groups. As the molecular markers, fibronectin, collagen I, and nestin were expressed upon the treatments. In addition, neuron-specific enolase, glial fibrillary acidic protein, and galactose cerebroside were also expressed [226]. On the other hand, in bovine aortic endothelial cells, GM2 and GM1 modified bFGF recognition to its receptor to inhibit the downstream signaling pathway of mitogenic protein-tyrosine phosphorylations such as PLC, MAPK, and protein-kinase-C [227]. The modulation of bFGF-induced mitogenesis by GM1 and GM2 seems to routinely occur in the serum of tumor patients. The result indicates that circulating GM1 or GM2 can regulate tumor growth because endothelial cell activation induces angiogenesis for tumor growth and metastasis. In the Schwann cells, bFGF also induces Schwann cell mitogenesis during myelin basic protein (MBP) promotion of cAMP-mediated Schwann cell growth. The MBP has therefore a mitogenic effect, which is promoted by binding with bFGFR. The MBP, which has a binding capacity to GM1, contains the cholera toxin B subunit homology. In addition, GM1 has been identified as the MBP and bFGF-binding receptors on the Schwann cells to stimulate Schwann cell mitosis [228]. In a recent study to explain the mechanistic action of gangliosides, the ganglioside-bFGF interaction to modulate the bFGF effects has been examined in the retinal Müller glial cells [229]. Treatment of the cells with GM1, GT1b, and asialoGM1 modified FGFR responses including mitogenesis, cell migration, and tyrosine phosphorylation of the FGFR with the cellular substrate influences. For example, gangliosides of GM1, GT1b, and asialoGM1 modified FGF-R autophosphorylation events in receptor tyrosine kinase signaling.

3.10 GM3-TGF-β Receptor Interaction

The biological event of epithelial-to-mesenchymal transition (EMT) exhibits a primary biological event observed in embryonic SC development as well as tumor invasion in epithelial cells. EMT is associated with development, wound healing, tissue fibrosis, cataract, mobility, and tumor invasion [230, 231]. TGF-β transduces EMT-elicited lens epithelial cells of humans after PCO. The formation of PCO of the lens epithelial cells is regulated by EGF, FGF, HGF, and TGF-β. In the anterior polar and posterior capsular cataracts, TGF-β is essential for epithelial–mesenchymal phenotype changes and extracellular matrix (ECM) production. The ECM includes ocular type I collagen and fibronectin [255–267]. GSLs involve in the EMT process. TGF-β, EMT, and GSL GM3 are associated with HLEs. GM3 induces the EMT in TGF-β1-derived HLE cells [232]. GM3 can regulate the TGF-β1-activated proliferation of glomerular mesangial cells [233, 234]. On the other hand, the gangliotetraosyl-Cer (Gg4) levels are reduced in the TGF-β-mediated EMT of mammary gland cells of normal murine together with the reduced level of UDP-Gal:β1,3-Gal-transferase-4 (β3GalT4) gene transcription [234–240]. In the TGF-β1-mediated EMT event, GM3 level is increased and correlated with the levels of migration and EMT signaling in lens epithelial cells of humans. The expression of GM3 synthase gene is transcriptionally upregulated by transcription factor Sp1 in the TGF-β1-treated HLE B-3 cells. GM3 recognizes the TGF-β1-activated TGFβ-R and participates in TGF-β1-driven EMT by interaction with TGFβRs. The two GSLs of GM3 and GM1 are largely present in cataract tissues of humans [241–244]. Caveolin-1 also upregulates the EMT event in the lens epithelial cells [245]. Two independent receptors of TGFβ-RI and TGFβ-RII embedded in PM lipid rafts activate MAPK when the EMT event is induced by TGF-β1 [246–249]. GM3 involves in EMT in the condition of TGF-β1 induction in HLE B-3 cells [250]. Phospho-Ser of TGFβ-Rs is found in the TGF-β1-activated HLE-B-3 cells. GM3 modulates the Ser-phosphorylation of TGFβ-RI/TGFβ-RII and Smad2/3, and also GM3 enhances binding of TGFβ-R in TGF-β1-derived HLE B-3 cells.

3.11 GM3-Insulin Receptor Interaction

As GSLs regulate epithelial differentiation and maturation on the cellular membrane, GM3 activates through tyrosine phosphorylation of the insulin receptor (IR), FGFR and PDGFR [198, 251, 252]. Ganglioside GM3 and IR are bound [252]. GM3 recognizes EGFR and consequently inhibits the EGFR Tyr phosphorylation, and also GM3 recognizes caveolin-1 in the caveolar microdomain on the PM of epithelial cells [253]. GM3 modulates the GFR signaling through the Ser/Thr phosphorylation. Moreover, in the adipocytes, the Ser phosphorylation of level IR is enhanced by the endogenously upregulated GM3 product in the TNF-α-elicited insulin resistance [251]. Similarly, GM3 has been known to associate with CD82 and caveolin-1 expressed in PM of epithelial cells and triggers the Thr^{654} phosphorylation, facilitating EGFR internalization [253]. The in vitro exogenous GM3 directly integrates and regulates IR.

GM3 level is surprisingly reported to be increased in insulin-resistant cells, and insulin IRS-1 phosphorylation is inhibited [254]. GSL synthesis inhibition reverses insulin signaling inhibition in the insulin-resistant cells. In the insulin receptor modulation of gangliosides [255, 256], the GM3's modulation capacity of insulin action is based on tissue or cell types. For example, GM3 is reported to promote insulin resistance on specific tissues such as adipose tissue. Nonetheless, the precise role of GM3 in insulin resistance is not clear in tissue or in vivo, although some of the in vitro activities were obtained. Therefore, further work is needed to mention the GM3 function in different tissues under pathological conditions such as obesity. In the current summary of G3 action, GM3 is reported to compete with caveolin proteins in PM for the competitive binding to the insulin receptor. The increased GM3 levels play a role in redistribution of insulin receptors in PM, or GM3 may promote Ser phosphorylation of IRS-1 to impair the insulin receptor signaling [257]. Thus, cell surface gangliosides are suggested to be the precise modulators in cellular responses upon physio-pathological conditions. In adipocytic 3T3-L1 cells, the accumulated expression of GM3 after TNF-α stimulation induced insulin resistance in obese condition, while enhanced accumulation of GM1 in aged conditions led to insulin resistance [258]. The TNF-α-induced GM3 expression eliminates the incorporated insulin receptors from microdomains and inhibits insulin receptor signaling. Such expressed GM3 blocks molecular recognition of the insulin receptor with Cav-1 caveola protein, causing the removal of the insulin receptor from membrane caveola and eventual dysfunction of insulin signaling. For the reverse demonstration of the GM3 effects, GluCer synthase inhibition counteracts TNF-a-induced pathogenic status mediated by insulin signaling, providing to normalized levels of GM2 and GM3. In Huh7 cells, the clustered GM2 was reported to inhibit IRr signaling by eliminating the IR from membrane microdomains of non-caveolar lipid rafts [259]. But, the fine molecular action is still unclear in insulin resistance. From the information described above, the gangliosides in action are not precisely generalized. It seems that monosialyl gangliosides such as GM3 also function as negative regulators, whereas oligosialyl or polysialyl gangliosides rather function as positive activators of the RTK signaling pathway. However, the action mechanism of gangliosides is not simply explainable for the distribution, composition, and structure of NeuAc residues of gangliosides [260]. As in case of GM1, GM1 accumulation influences on the fate of cell surface expression of microdomain raft-linked membrane proteins and reduces membrane fluidity and unbalanced signal transduction [261].

3.12 GM3/Gb3/GD3-Fas (CD95) Receptor Interaction

Plasma membranes contain specific clusters named lipid rafts. The lipid rafts or so-called microdomain islands on fluid mosaic structures are mainly composed of hydrophobic lipids and hydrophilic proteins. Such lipids are comprised of sphingolipids including gangliosides and sphingomyelin and cholesterol, while proteins are such as membrane receptor proteins. They function as active traffic and cell remodeling actors. Apart from the positive signaling aspects after internalization, lipid rafts are also associated with negative signaling aspects such as cell death event. The negative case is seen in the fact that lipid rafts are suggested to function in receptor-mediated T cells apoptosis. Such roles of membrane-formed lipid rafts for apoptotic induction have been mostly elucidated in T cells, as the T cells apoptosis has well been known to undergo through CD95/Fas signaling. The event requires the receptor engagement recruiting CD95/Fas [262, 263] and TNF-family receptors [264] to cellular PM lipid rafts and proapoptotic bcl-2 family proteins to mitochondrial membrane.

Ceramide in specific lymphoid cells induces apoptosis or programmed cell death signaling by rapid accumulation upon CD95 interaction, following the sphingomyelin hydrolysis by acidic sphingomyelinase. Some gangliosides including GM1 [265], GM3 [266], and GD3 [267–270] are also known to induce apoptotic cell death. GD3 and GD1b elicit apoptotic cell death in human Colo-205 colon cancer and breast carcinoma SKBR3 cells. GD3 induces apoptosis of hepatocytes and hematopoietic cells. GM3 exhibits the induction of apoptotic cell death of human glioblastoma cells which are features of high-grade invasiveness. GM3 is predominantly present in GSL-associated microdomain of cellular membranes in T cells. CD95/Fas interaction induces a GM3-caspase-8 association. GM3 and GM1 are associated with caspase-8 upon CD95/Fas binding (Fig. 3.8). CD95-mediated apoptotic signal increases the level of ceramide synthesis and consequently rapid increase in GD3 synthesis is followed in hematopoietic cells. Such effects are not observed by certain gangliosides including GM1, GD1a, or GT1b. Among gangliosides, GD3 is specifically associated with the Fas/CD95-regulated apoptotic pathway in the Fas-induced Jurkat T cells via colocalization with membrane lipid rafts so

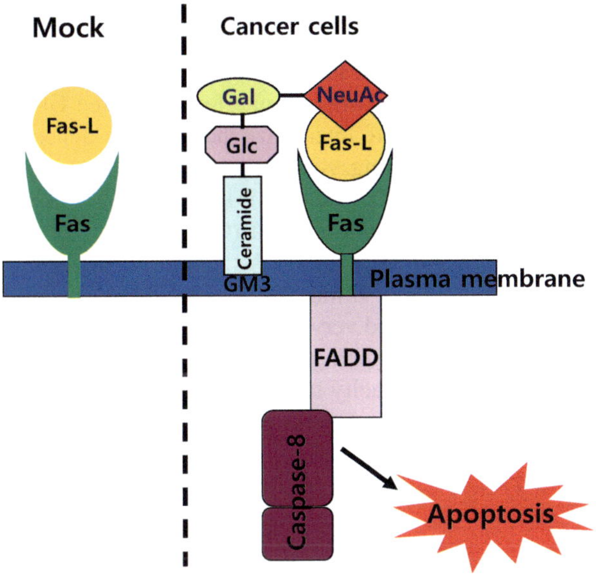

Fig. 3.8 Apoptotic cell death through GM3-Fas-L/Fas binding in human colon cancer. Membrane-anchored GM3 recognizes with Fas-L and consequently induces the GM3/Fas-L and Fas receptor interaction to recruit adaptor molecules of FADD and caspase-8. The terminal NeuAc residue of GM3 binds to Fas-L

that GD3 induces the transcription factor genes to elevate apoptosis. This indicates that ganglioside GD3 expression and redistribution contribute to mitochondrial alterations leading to cell death [271, 272].

As the GSL-glycans-receptor interaction for the PSI type, the molecular interaction between the CD95/APO-1/TNFRSF6 or Fas receptor and globoside Gb3 is another case. Fas receptor's carbohydrate-recognizing domain recognizes Gb3 and LacCer with its specificity. However, it does not interact with Gb4 or any other gangliosides [22]. Fas receptor is a family of TNF receptor superfamilies, which is an integral activator of apoptosis of cells. The Fas receptor recognition to its ligand FasL induces apoptotic death signaling through the recruitment of initiating proteins of procaspases-8 and Fas-associated death domain (FADD), the caspase autoactivation, and resulting apoptosis. In other way, Fas is also involved in the induction of non-apoptotic behaviors such as tissue proliferation and regeneration. A conserved extracellular motif for GSL binding of Fas regulates the internalization and ligand binding signaling. This binding motif function of Fas is archived by clathrin-mediated internalization, triggering induction of its cell death signals. If the binding motif loses the roles, the stimulated receptor needs an alternative pathway of endocytic internalization that is clathrin-independent, but cholesterol-dependent lipid rafts as well as ezrin-dependent [273]. This is a mechanism how does it extinguish its cell death signaling and promote its non-death activity in the multifunctional Fas receptor-ligand system. The PCI event in the Fas-GSLs interaction has functional consequences because the carbohydrate-recognizing domain of Fas designates its internalization pathway, as well as the signaling upon FasL-Fas interaction.

Gb3 is generated from LacCer by α1,4-Gal-transferase (Gb3/CD77 synthase). GM3 synthase (ST3GalV) catalyzes competitively to Gb3 synthase due to their common precursor usage of LacCer. When interacted with Gb3, the Gb3-bound Fas undergoes its internalization through clathrin-associated endocytosis, triggering the signaling of a cell death and activation of procaspase 8 to caspase-8 as well as caspase-3. The blocking and interference of the Gb3-Fas receptor recognition promotes Fas receptor to internalize though ezrin-dependent endocytosis and activate a proliferation signaling via downstream MAPK signaling cascade [22]. The extracellular domain of Fas receptor is bound strongly to the LacCer and Gb3, but weakly to GD3. As GSL-binding motif (GBM) is the site of the recognition, the essential peptide of the GBM of mouse Fas receptor (GBM-mFas) has been confirmed to bind to lipid monolayers. Similarly, the peptide has been found to be bound strongly to Gb3 and LacCer, and very weakly with Gb4 and GD3. However, no interaction was observed with the sphingomyelin (SM) that is deficient for a sugar head group. Considering its capacity to bind proteins, it is noted that Gb3 is also a non-protein receptor responsible for Shiga toxins (STx), which are known as an exotoxin produced by *Shigella dysenteriae* and enterohemorrhagic *Escherichia coli* (EHEC) [274, 275].

3.13 Lyso Gb3-Notch Ligand Delta-Like 1 Interaction

During developmental regulation of GSLs, Notch ligand Delta-like 1 (Dll1) of mammals is another model, where GSL carbohydrate binding domain is embedded in the ligand itself [53]. Notch receptors are heterodimers [276] with an extracellular domain of N-terminal region, linked to C-terminal membrane-anchored domain in a noncovalent binding way. In the process of recognition of Notch receptor with its ligand of Delta-Serrate-Lag-2 (DSL), the Notch receptors are proteolytically cleaved and this event leads to transcriptional upregulation of Notch target genes. Notch signaling is activated by a tightly regulated mechanism in the signal-emitting cells or signal-receiving cells [277]. The Notch signaling is associated with endocytosis and endosomal trafficking [278]. In Drosophila *Caenorhabditis elegans*, the extracellular domain in Drosophila Delta structural motif corresponds to the GSL-binding motif to interact with GSLs [279]. The Dll1 trafficking and GSL-binding motif in mammalian Dll1 are closely related. GSL-binding motif is located on Dll1 and GSL-binding motif is essential for Dll1 trafficking via prevention of degradation and shedding. GSL-binding motif is basically found in Drosophila Notch ligands and they are crucial for Dll1 endocytosis. Point mutation of its conserved Trp residue leads to the rapid inactivation of Dll1 molecule, without recycling to the cell surface and Notch signaling activation. During ligand-receptor recognition and endocytic internalization of the binding complex, the receptor cleavage activates the kinase signaling in the signal-receiving cells [53]. If the key residues of the Dll1-carbohydrate binding domain are mutated, Dll1 is inactivated by degradation and recycling inability to the surface of cells and thus to stimulate the Notch signaling. Blocking the GSLs expression in the signal-emitting cells impairs the related Notch signal stimulation [279]. Therefore, the GSLs-Dll1 interaction as the PCI is crucial for the Notch signaling, indicating that GSLs are the interacting place to accumulate Dll1 in PM lipid-rafts to activate the Notch signaling and upregulate its binding affinity for the Notch receptor [35, 90]. Silencing in the GSL synthesis genes inhibits Notch signaling activation. GSLs interact with Dll1 and function as co-factors to activate its physiological activity. In the podocyte injury of the Fabry nephropathy, Lyso-Gb3 accumulates in serum. The Notch1 signaling as a podocyte injury mediator contributes to lyso-Gb3-stimulated cellular responses in human podocytes during cultivation. Lyso-Gb3 stimulates Notch1 signaling of podocytes and increases Notch targets of Notch1 and HES1, which are canonical transcriptional targets. Lyso-Gb3 induces Notch1-elicited inflammatory response and Fabry nephropathy [280]. Further comprehensive effects to the GSL-dependent signaling regulation is dependent on the heterogeneous diversity of ceramide backbone structures and the bound glycan moieties because the GSL-ceramide backbone directly interacts with transducer molecules [281, 282]. Cholesterol and GSL-hydrophobic portion form systemic membrane domains to operate the clustered and distributed receptors.

3.14 GM1-GPCR Interaction

GSLs are classified as minor essential components in eukaryotic cellular membranes with multiple cellular events. GPCRs as transmembrane signaling molecules [283] regulate multiple biological events and are valuable targets for therapeutic drug exploitation [284]. Interestingly, over 50% of the currently available drugs are known to regulate GPCR activity [285]. The core structure of GPCRs is linked to extracellular and intracellular loops. Known ligands bind to the central core region of GPCRs or the extracellular loop regions of GPCRs. The intracellular loop 3 is linked to G protein and β-arrestin, which are coupling to exert GPCR function. As several membrane receptor proteins such as GPCRs functionally rely on sphingolipids, depletion of GSLa affects receptor function [286, 287] by direct recognition or indirect recognition, or their combination. Sphingolipids occupy about 20% of total membrane lipids, functioning in cellular signaling, cell growth, cell differentiation, and neoplasmic transformation. Direct interactions are based on the concept of the consensus "sphingolipid binding domain" (SBD). In literature, SBD was initially reported to be characterized in gp120 glycoprotein of HIV-1 surface envelope, certain receptors, amyloid proteins, bacterial toxins, and infectious viral adhesion proteins [288]. The SBD motif is known as the LNKWTLGQVTC, which consists of the conserved basic, aromatic, and turn-making amino acids. The aromatic amino acids interact with the sugar moiety of GSLs. The most well-known sphingolipid-binding GPCR is the serotonin1A receptor [286], which plays a role in cognition, behavior, and the development of brain.

GM1 binds to proteins to modulate GPCRs, ion transport, neuronal functions, and immune responses [133]. Other gangliosides of GT1b, GD1b, GD1a, GQ1c, and GT1a are additionally sialylated to GM1 structures (Fig. 3.9). Interestingly, such

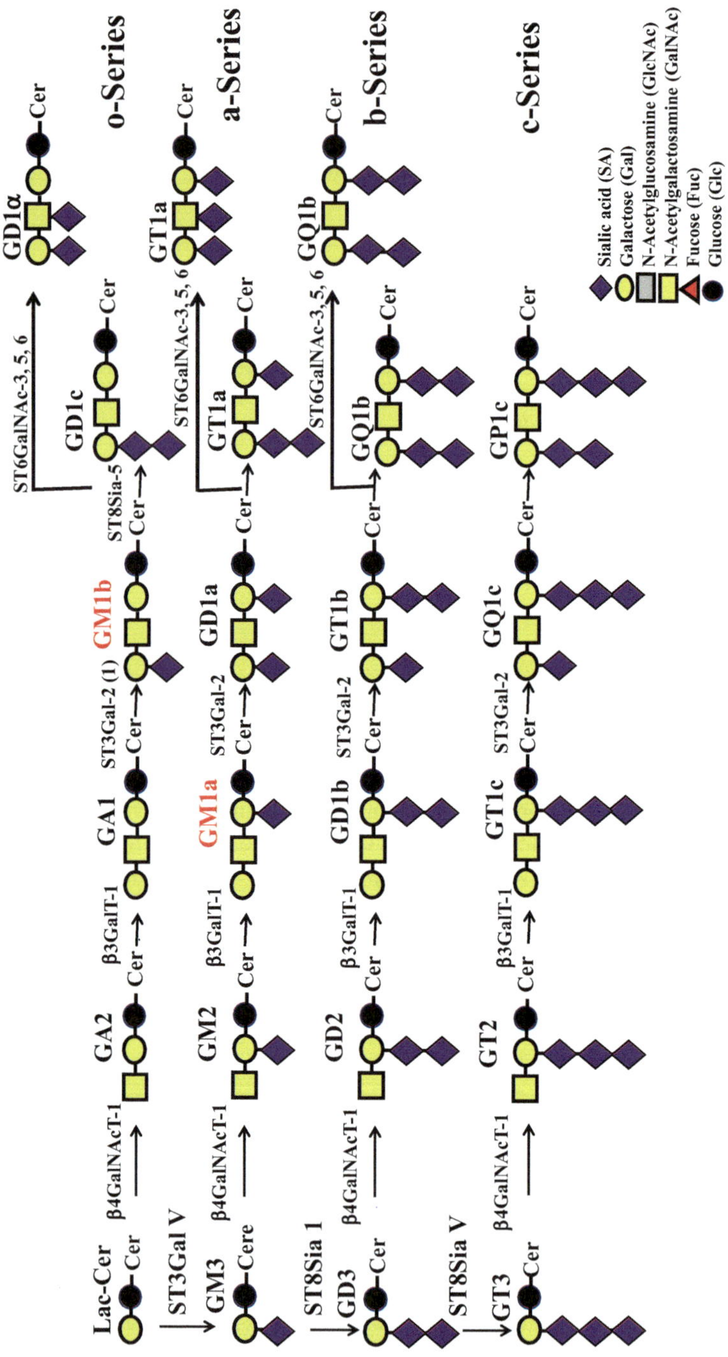

Fig. 3.9 Ganglioside biosynthetic pathway of o, a, b, and c ganglio-series

additional SA residues are easily removed by enzymatic digestion by membrane neuraminidase (sialidase-2 or Neu-2) [289], while the SA residue of GM1 is not sensitive to mammalian sialidase digestion. Therefore, sialidase enzymes elevate GM1 amounts if GM1 ganglioside is essential for the function. GM1 also modulates opioid activity through opioid receptor interaction with prototypic GPCRs. The μ- and κ-type opioid receptors are similarly regulated. The excitatory Ca^{2+} influx is blocked by GM1-specific CtxB [290], as demonstrated from inhibitory action to excitatory action of the δ-opioid receptor-expressing cells [291]. The carboxyl group of SA negative charge is required, as demonstrated by mutagenesis technology targeting the specific sites of the receptor gene of δ-opioid replaced from amino acid at the Arg-192 to Ala residue. The Arg to Ala mutation prevented the GM1-driven excitatory response for charge-charge interaction [292]. GM1 also regulates adenylyl cyclase activity associated with GPCR like β1-adrenergic receptor (AR) [293]. However, GM1 has no effect on another type of the β2-AR. GM1 also stimulates prostaglandin E1 (PGE1)-elicited cAMP generation in Neuro-2a cells [294].

Typically, GM1 influences the GPCRs function as a neurotransmitter receptor. GM1 recognizes the SBD of the serotonin1A receptor via the extracellular loop 1 location. The SBD motif possesses specific amino acids including the evolutionarily conserved amino acids sequence in the serotonin1A receptor [295]. SBD motif of the serotonin1A receptor interacts primarily with gangliosides [296]. The SBD site-GM1 binding stabilizes a "flip-out" stereochemical conformation position in which amino acid W102 of the extracellular loop 1 flips out from the receptor central lumen region for the membrane [297]. Direct binding of GM1 to the SBD site of the serotonin1A receptor occurs in vivo and implies malfunctioning of neuronal GPCRs under neurodegenerative diseases. GM1 interaction with the serotonin1A receptor SBD in extracellular loop 1 is interested because the SBD motif sequence is well conserved during evolution in the serotonin1A receptor. The direct recognition of GM1 to the SBD sequence of the serotonin1A receptor is meaningful of validation because it occurs in vivo.

As another type of GPCR, GPR37 is linked to the pathology of Parkinson's disease during intracellular accumulation or aggregation of GPR37. It is used as the E3 ubiquitin ligase substrate, named parkin [298]. The GPR37 interaction with GM1-containing lipid rafts protects from intracellular GPR37 aggregates toxic and specific for parkinsonism. Inhibition of extracellular prosaposin which named sphingolipid activator protein-1, PSAP, or sulfated glycoprotein-1 destructs GPR37 binding to GM1-associated lipid raft microdomains and consequently stimulates its intracellular disposition and accumulation. The functional complex formation between GM1, GPR37, and PSAP in the PM protects cells from intracellular accumulation of GPR37. GM1 may prevent the PD progress through stabilization of GPR37 in the PM [299]. If GPCRs are activated prolongedly or repeatedly with agonists, the desensitization and internalization are evoked, allowing the blocked responsiveness to agonist activation. Recently, certain gangliosides are known to influence B2 bradykinin receptors, B2Rs, and several gangliosides effectively internalized B2Rs [154].

3.15 GM1-Human Serotonin Receptor Interaction

The localization and function of several GPCRs are modulated by PM-associated lipid rafts microdomain that are formed with cholesterol, gangliosides, caveolin, and sphingolipids. 5-hydroxytryptamine (5-HT) or serotonin functions through currently known 14 different receptors of serotonin, regulating metabolism, mood, and contraction of smooth muscles. Among serotonin receptors, serotonin1A receptors also belong to one group of the GPCRs, functioning in the behavior, cognitive, and developmental events. With regard to the serotonin1A receptor (SR1A), GSLs are required for the SR1A activity and the SR1A preferentially binds to the "GSL-binding region" in the SR1A. The distribution of GSLs in the bilayer with cholesterol collaborates with several membrane proteins such as GPCRs. One of the GSLs-dependent examples of GPCRs is the serotonin1A receptor [286]. Experimental metabolic depletion of GSLs induces changes in the receptor function [300]. GSLs effect on GPCR function has been attributed to direct interaction, as several GSL-associating PM proteins have a conserved region named "GSL-binding domain." GSL-binding domain is also explained in a gp120 of HIV-1 envelope protein, Aβ-proteins, toxins, or virus proteins. The GSL-binding domain motif is characteristic for aromatic and basic amino acid residues. The GSL-binding motif in serotonin1A receptor is also evolutionarily conserved [295]. The GSL-binding domain of the serotonin1A receptor prefers gangliosides to other GSLs [296]. The ganglioside GM1 binds to the extracellular loop 1 region in the serotonin1A receptor as the GSL-binding site, regulating the ligand binding and receptor function [301].

3.16 GD3-α1-Adrenergic Receptor (AR)/Transglutaminase 2 Interaction

GD3, a di-SA-containing glycosphingolipid, is reported to recruit membrane transglutaminase 2 (TG2) in K562 cells, known as human chronic myelogenous leukemia (CML) cells [302]. The α1AR-TG2 complex signaling pathway upregulates GD3 roles to erythrocytic differentiation. ARs have three sub-forms of α1AR, α2AR, and βAR. α1AR has also 3 subgroups of α1A AR (or called C), α1B AR, and α1D AR, classified by their ligand-specific receptor recognition and activation signaling events [303]. Naturally discovered α1-AR ligands include catecholamines, epinephrine, and norepinephrine. ARs also couple with TG2, where TG2 has a Ca^{2+}-dependent transamidase. GTP-bound TG2 is a membrane receptor that the α1-AR stimulates GTP-recognizing capacity of TG2. TG2 highly accelerates rapid erythroid differentiation process of K562 cells [304].

Epinephrine, a representatively known AR agonist, membrane-recruits TG2 with GTP activity as well as stimulates GD3 expression via PI3K and Akt activation as well as GTPase activity in the TG2 activates Akt activity. However, the precise molecular mechanism(s) how gangliosides lead to cellular differentiation is not well

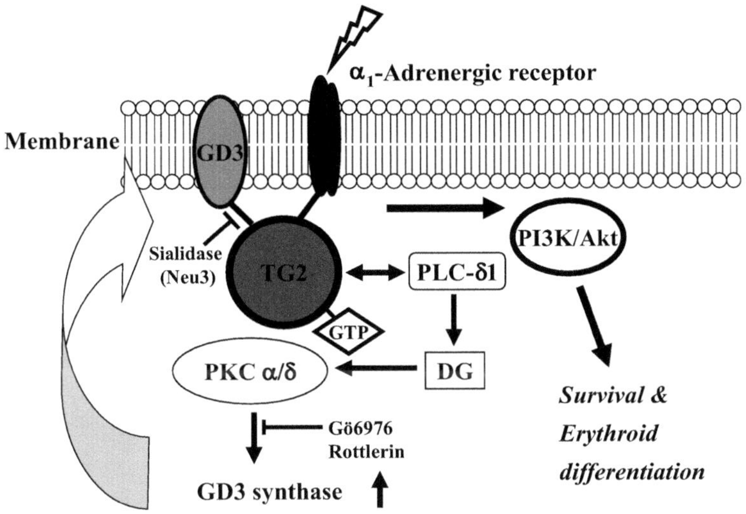

Fig. 3.10 GD3-α1AR/transglutaminase 2 interaction to induce erythrocytic differentiation. (Adapted from Ref. 235. Ha SH et al. (2018) Oncotarget 8(42), 72,205–72,219)

explained. CMP-NeuAc:GM3 α2,8-ST as the GD3 synthase is frequently abbreviated to ST8Sia-I or SAT-II enzyme. This enzyme catalyzes the synthetic pathway of GM3 into GD3. GD3 is an actor of cell differentiation; however, tightly regulated GD3 function affects its distribution, localization, structure, and expression [305, 306]. The TG2 and GD3 axis is specific for α1-AR. The gene expression of α1AR-TG2-PKCsαδ-GD3 synthase axis is coupled together. α1AR-TG2-GD3 production is therefore required for K562 leukemic differentiation to erythrocytes. The key biological point was that GD3 directly binds to α1AR-TG2 to induce GD3-α1AR-TG2 erythrocytic differentiation, concluding that GD3 is an erythroid differentiation regulator in CML cells (Fig. 3.10). This evidence allows a therapeutic medium for leukemia possibility.

3.17 GT1b-Amyloid-β (Aβ)-Derived GSL-Binding Domain Interaction

In a recent study, $A\beta_{1-25}$ has been shown to have SBD peptide [307]. For the direct evidence, the fluorescently tagged $A\beta_{1-25}$ has been demonstrated to act as a probe SBD. The $A\beta_{1-25}$ SBD tag exhibited a high binding affinity for GT1b recognition, compared to GM1. Cholesterol enhances the binding capacity of Aβ SBD peptide with GM1 and this enhanced binding has been attributed to conformation shift of gangliosides with hydrogen bonds through CH-π interactions [308]. However, the molecular recognition of tagged $A\beta_{1-25}$ SBD and poly-sialyl GT1b is not clear. The $A\beta_{1-25}$-derived SBD is also another case as the PSI, although the molecular interaction mechanism between $A\beta_{1-25}$ SBD and molecules associated PM domains is not clearly explained. The question how the $A\beta_{1-25}$ SBD motif recognizes and binds the PM lipid surfaces is recently answered by the result that $A\beta_{1-25}$ SBD adopts a coiled helix-coiled structural motif and this conformation-shifted motif efficiently binds to multiple embedded GT1b through CH-π recognitions and salt bridges. Therefore, chemically generated CH-π interactions and electrostatic driving forces formed by the SBD binding to GT1b have been considered to be the major driving forces for the recognition. It is largely suggested that glycolipid binding domain is also structurally conserved to some sphingolipid-binding proteins such as Aβ, extending to other HIV gp120 and prion proteins [309].

3.18 GM1-AMPA Ionotropic Glutamate Receptor (AMPAR) Interaction

Gangliosides as SA-bearing GSLs are cell determinants, and the ceramide moieties are embedded in the PM extracellular leaflet and the carbohydrates are directed to extracellular (outward) regions. The ganglioside biosynthesis defects display congenital disorders such as intellectual disability associated with seizure susceptibility, learning disability, and dysregulation of excitatory neurotransmission [310]. Congenital disorders found in humans are associated with the gangliosides for learning, memory, and excitatory neurotransmission. Ganglioside-recognizing proteins such as ganglioside-mediated AMPAR trafficking are thus involved in human congenital disorders. For example, Glu, a known brain excitatory neurotransmitter of mammals, recognizes the AMPA glutamate receptors (AMPARs or GluR AMPAR) to exert excitatory neurotransmission. The AMPAR is mainly expressed at postsynaptic membranes and is involved in excitatory neurotransmission that contributed to synaptic plasticity, learning, and memory [311]. Therefore, AMPARs at synapses are tightly regulated. Using ganglioside capture methods, the GT1b-binding proteins such as Thorase have previously been captured for neurotransmitter receptor trafficking. For example, Thorase is an ATPase-dependent endocytosis regulator of the GluR2 AMPAR subunits as an AMPAR ion channel activator [312]. GluR2 does not bind to GT1b, but specifically binds to GM1. Noncleavable ATP (ATPγS) treatment disrupts ganglioside binding to enhance the GluR2 and receptor-trafficking protein association, indicating the gangliosides sequester GluR2-bearing AMPARs and GluR2-delivery complexes in a reversible and ATPase-requiring way, and this consequently modulates GluR2-bearing AMPAR endocytosis, permeability of AMPAR ion channels, memory and learning ability and synaptic plasticity [313].

The GluR2-bearing AMPARs bind to GM1 only, whereas Thorase, γ-SNAP, NSF, and Nicalin-bearing AMPAR-trafficking complexes (ATCs) bind to GT1b. The protein-ganglioside interactions are abolished by ATPγS, whereas direct binding of GluR2-bearing AMPARs and ATCs is promoted by ATPγS. The GluR2-bearing AMPARs and ATCs are sequestered by GM1 and GT1b, respectively.

AMPARs as transmembrane proteins can directly interact with GM1. Colocalization of GluR2 and GM1 in synaptosomes supports the direct AMPARs-GM1 binding [314]. The receptor-ganglioside interaction has been demonstrated as affinity-captured ganglioside-binding proteins by proteomic mass spectrometry. Such captured brain proteins specifically bind to the GT1b (GT1b > GM1) and regulate neurotransmitter receptor trafficking. GluR2-bearing AMPARs do not recognize GT1b but binds to GM1. Treatment with ATPγS reduces the ganglioside GT1b-binding capacity, but increases the levels of AMPAR recognition to NSF, Thorase, and Nicalin. GT1b-lacking mice express higher brain level of Thorase, while Thorase-lacking mice produce the increased GT1b expression level. The hippocampus neurons treated with neuraminidase enzyme cleave GT1b-attached SA residues off and thus significantly reduces the size of GluR2 puncta in cell surface. These facts proposed a possible mechanistic model, suggesting that GM1-bound GluR2-bearing AMPARs are actively separated from ATCs bound with GT1b. Release of the bound GT1b gangliosides stimulates GluR2-AMPAR interaction with its transportation complex, ATCs, enhancing endocytosis capacity. Lack of ganglioside biosynthesis also reduces the expression of synaptic GluR2-bearing AMPARs, consequently leading to intellectual defects and increased susceptibility to seizure, because AMPARs regulate seizure phenotype [315] and AMPAR trafficking regulates learning and memory [310]. AMPARs are classically a group of tetrameric transmembrane proteins family with a dimer of dimers bound with four GluR1–GluR4 subunits [316]. Among them, a key subunit GluR2' trafficking and expression is highly regulated, where the brain gangliosides regulate the trafficking of GluR2-bearing AMPARs. Therefore, the human congenital disorders caused by defected ganglioside biosynthesis are known to contribute to AMPAR disruption associated with seizures and cognitive deficits.

Chapter 4
Viral Protein Interaction with Host Cells GSLs

Viruses attach and adhere surface receptors to target host cells, and the receptors determine viral tropism of hosts and tissues. Cell surface glycans are used as the virus receptors. The enveloping viruses generally recognize surface receptors of hosts as the initial step in the viral infection cycle to host and further define the viral host range determinant. From the linkage diversity and carbohydrate structure motifs, which are apparently distinct from proteins, most pathogens or viral agents use glycans as infection receptors. The molecular structure of the virus receptors is topical in the field of virus research, but no concrete consensus logics have emerged yet. The viral interaction with the host cell surface is rather the complexed phenomenon with a multiple stage. Such multiple processes possess multiple recognitions and interactions with multiple cell surface molecules as well as accompanying conformational shifts in the virus-produced proteins. In the enveloping viruses, the adhesion-strengthening attachment strategy is affordable for virus-host receptor recognition and interactions to give higher affinities between the receptor and ligand interaction. In that meaning, gangliosides may function in virus-host cell interaction as primary receptors or co-receptors. The co-receptor concept has well been exampled in the section HIV infection.

Still independent investigations implicated many candidates such as gangliosides, integrins, and other membrane proteins for viral binding, attachment, and entry into host cells. The SA-containing glycans as attachment receptors have been elucidated for a broad range of DNA/RNA viruses and enveloping and nonenveloping virions. Because SAs are present on all the cells in vertebrates and sialylation is a terminator of glycosylation in lipids and proteins, sialic acids are frequently recognized as physiological candidates for host cell attachment via target receptors. For example, numerous pathogens are reported to attach to SA-linked receptors to penetrate cells, as in examples of various viruses and pathogenic bacterial exotoxins. Although multiple pathogens including viruses, parasites, and bacteria recognize SA residues on molecules in host cell PMs, certain pathogens cause tumors, too. Among SAs, the well-known binding counter is a NeuAc form, while NeuGc form and 9-O-acetylated SA residues are also used as recognition receptors. Viral pathogens which

C.-H. Kim, *Glycosphingolipids Signaling*,
https://doi.org/10.1007/978-981-15-5807-8_4

recognize gangliosides as receptors are known for many infectious viruses including simian virus SV40, influenza virus, and polyomavirus. Some bacterial pathogens are also known to recognize gangliosides and the actual binding components are bacterial toxins and adhesins as well as the SA-binding adhesin from the *Helicobacter pylori.* Apart from the pathogens, upon Singlec-7 binding, gangliosides are also known to modulate immune-related cells like NK cell cytotoxicity.

The viral attachment, recognition, interaction, and adhesion to the host receptors expressed on the cell surfaces determine the fate of the virus, because such interaction contributes to virus attachment, adhesion, penetration, entry, and infection to the host cells or some cases contribute to the destruction of cytosolic region in the infected virus. Carbohydrate interactions are precisely controlled in the viral life cycles in attachment on the cell surface and cellular trafficking [317]. Virus attachment and entry are multistep process including binding to the cells and attachment receptor for internalization. Carbohydrate moieties of host cell surfaces are used to recognize and infect epithelial cells. In most instances, among carbohydrates, SA and SA linkages determine virus host range, host specificity, tissue tropism, host cell propagation, and pathogenesis, where most SA receptors consist of terminal SA linked to the penultimate Gal or GalNAc by a SAα2,3 or SAα2,6 linkage. With these biological functions of SA, gangliosides are the better receptors than SA alone for many viruses due to the longer lengths of glycan chains. Viruses are regularly subclassified into multiple families, depending on their nucleic acid presence and content with RNA virus or DNA virus, capsid symmetrical conformation with helical virus, icosahedral virus, or complex virus, and the lipid envelope existence with enveloping virus or nonenveloping virus. Each specific characteristic displays each distinct replication method, even though all they belong to obligate intracellular viruses that absolutely depend on the intracellular machineries in their life cycle. Viral infection into their host cells is multiply processed with cell surface receptor binding, PM fusion for enveloping viruses, endocytosis, membrane internalization or penetration for enveloping and nonenveloping viruses, replication site delivery and replication and virion progeny production. Virus recognition and binding an attachment to the host cell surfaces are the initial steps and events for their host infection. In virus cases, they infect their target cells of susceptible hosts by attachment of the virus. Such interaction is therefore a valuable target for antiviral therapy. Host cell receptors expressed on the PM surfaces adhere viruses. The receptors function as determinants of viral infection and tropism. Receptors are flexible for attachment, host decision, and infection to specific cells as host. For some viral families, large progress on understanding the molecular mechanism involved in viral penetration and entry into host cells has been made. Viral infection needs its entry to host cells, which consist of virus recognition, binding, and attachment to the host cell surfaces. Thereafter, the virus penetrates into the interior membranous side and is followed with the next viral disassembly of the capsid proteins of virus in host cells.

In the host cells, the genetic amplification and propagation of viruses is progressed in a programmatic way. Such serially continued steps are crucial for viral agents in trafficking to the extracellular environmental matrix for the cytoplasmic compartments, where viral multiplication takes place in the host cells. Viral penetration and entry mechanisms are also adapted to express viral diseases and pathogenesis as the events frequently direct selective determination of target cells within the host toward the virus-induced disease site. In addition, penetration-entry steps trigger signaling pathways to influence the host cells antiviral state or apoptosis. Viral attachment is monophasic and multiple receptors are involved, depending on adhesion strength. Viruses are capable of binding different cell surface molecules such as lipids, proteins, and carbohydrates. Some groups are just simply attachment-related factors, concentrating virus on the cell surfaces. But other groups are used as direct accessible co-receptors or receptors to mediate host attachment of virus and penetration to host cells. Some viruses alternatively utilize certain receptors with a cell type specificity. Virus-binding molecules on the cell surface are typically glycoconjugates such as GSLs, glycoproteins, and proteoglycans [318].

In the cells, most glycan portions are extended to the extracellular area. Therefore, viruses cannot interact with the distal membrane apical glycans. Thus, viruses bind to the proximally located protein cores of PM of host cells. In general, a virus interacts with an attachment molecule named receptor, generally a carbohydrate glycan for adherence via low-affinity recognition and recognition to the cell surface, indicating an additional receptor is required for stable recognition with high binding affinity. In most cases, glycan binding is displayed by electrostatic forces to anionic charged SA residues-carrying glycan moieties or glycosaminoglycans. In rare cases, the glycan moieties are attached on both glycoproteins and glycolipids. Some viruses just bind to cell surface glycans only to facilitate viral entry, although the relationship between the glycan structures and status of viral disease incidence is incompletely explained. Virus recognition and interaction with cell surface receptors frequently enhance intracellular signaling cascades to enhance virus entry potential [319]. The first step of infection of virus to host cells is attachment of viral particles. The cell surface receptors are targeted by viruses, depending on cell type, and receptor specificity, as expressed for host cell tropism, specificity, pathogenic incidence, and virulent direction. Therefore, the knowledge how a virus recognizes, binds to, and interacts with their typical receptors is a primary step to prevent virus infection and spread.

Many enveloped viral pathogens decorate their surfaces with glycans. Examples of **envelope proteins are** the envelope glycoprotein (Env) of HIV-1, HA of influenza virus, coronavirus spike (S) glycoprotein, Ebola virus glycoprotein, Lassa virus glycoprotein complex (GPC), and flavivirus envelope (E) glycoprotein of dengue and Zika (Fig. 4.1). On the other hand, secreted viral proteins are also glycosylated. Examples include the nonstructural protein-1 (NS1) of flaviviruses, the secreted GP of Ebola, and the secreted glycoprotein G of HSV. To date, eight different virus families are known to use sialoglycoconjugates for attachment. They are enveloping and nonenveloping viruses as well as RNA and DNA viruses in structures. The eight viruses include Adenoviridae, Coronaviridae, Orthomyxoviridae, Paramyxoviridae, Picornaviridae, Parvoviridae, Papovaviridae, and Reoviridae. Among them, many viruses including the adeno-associated bovine virus, human parainfluenza virus as respirovirus, influenza virus, murine norovirus, Newcastle disease virus,

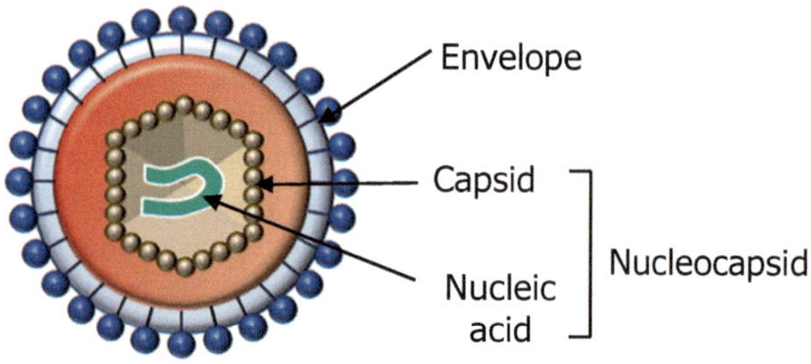

Fig. 4.1 Structure of enveloped virus

paramyxoviruses, Polyomaviridae family, Rabies virus, rotavirus, and Sendai virus [320–338] use gangliosides as receptors. Several viruses including influenza virus are known to use carbohydrate glycans as a principal receptor [339], whereas some other viruses including herpes simplex virus [340] and reovirus [341] use carbohydrate glycans as just initially adhesive molecules prior to binding specific molecular receptor for the real attachment, which is called adhesion strengthening event. Virus-carbohydrate recognition designates cell type susceptibility; however, the detailed information of each glycan to viral infection is very limited for most cases of carbohydrate-recognition and binding viruses.

4.1 GM3/Gb3-HIV Infection

In 1981 acquired immunodeficiency syndrome (AIDS) was first reported [342] (caused by HIV). The HIV genomic material comprises two plus (+) sense single RNA strands. Viral envelope contains the glycoprotein, gp120. HIV virions enter their host cells by viral host-cell membrane fusion [343]. HIV expresses a coat adhesin protein, gp120, which is composed of trimeric proteins with their high glycosylation. The virus-borne gp120 protein interacts with the CD4 protein expressed on Th cells. In addition, the gp120 protein has been known to recognize chemokine co-receptors, which are named CCR5 for R5 HIV1 strains and named CXCR4 in the X4 HIV1 strains with heterogeneity. The gp120 recognizes GalCer, which is the first identified, and also GalCer 3′ sulfate ester named sulfatide [344]. GalCer seems to act as an alternative or replacing receptor in CD4-deficient cells such as neural and GI epithelial cells [345] or reproductive epithelium. This raises a question whether epithelial cells are actually infected by HIV. However, if not, such cells can probably serve as a reservoir or latent cells for viral DNA for a longer latency. In fact, GalCer mimics prevent HIV infection to both CD4 positive and negative cells. GM3, GD3, and Gb_3 are bound to gp120. The gp120s expressed from dual tropic phenotype of R5X4 HIV strains recognize specifically GM3, while gp120s expressed from the X4 HIV strains bind predominantly to Gb3.

In view of GSLs expressed on viral membrane, HIV-1 interacts with DCs and is disseminated to CD4+ T cells through so-called trans-infection pathway by virion membrane incorporation of the GSLs such as GM3 produced by host cells. This

DC-driven trans-infection pathway contributes to multiple CD4+ T cell infection, as the mechanism has been appreciated. DC-ligand interaction elicits type I IFN signaling to activate DC-enhanced T cell trans-infection. The type I IFN-elicited Siglec-1, named CD169, acts as the DC receptor to GM3-dependently capture HIV. DCs capture HIV-1 membrane incorporation of the α-2,3-SA gangliosides. The candidates are GM1 and GM3 that have α-2,3-SA residues, but GM3 is effective for viral particle capture. Virions are captured through Siglec-1, CD169 on the PM. GM3 depletion from viral membranes or Siglec-1 depletion from DCs lost HIV-1 activity of capture and internalization as well as T cell trans-infection by DCs. Siglec-1 on macrophages captures murine leukemia virus [346] and HIV [347]. In the case of MLV, Siglec-1-mediated capture by macrophages is followed by migration to lymphoid follicles and trans-infection of B cells. Siglec-1 drives virion capture in HIV-infected macrophages. Siglec-1 recognition captures virions and forms the virus particles. Siglec-1 captures virus particles because it attaches to gangliosides on the virus envelope. Therefore, Siglec-1 is also an important receptor in retroviral particle capture and transmission. Thus, pathogen parasitization of host produced attachment tool like the GM3-Siglec-1 interaction enables DC-mediated HIV dissemination.

On the other hand, classically, the viral envelope gp120 first recognizes its primary receptor on host cells, CD4, using a binding motif in its second constant (C2) region with a conformational shift in HIV gp120 which exposes its third variable loop (V3) which contains a consensus motif that binds to a seven transmembrane-spanning chemokine co-receptor [348]. HIV-1 infection and cell susceptibility are still unclear, because the receptor functions of GSLs during HIV infection are rather complex. HIV recognizes and interacts with the GSLs as viral receptor expressed on the cell surfaces of hosts when the host cells are negative for expression of the canonical receptor CD4. GSLs that gp120 recognition domain recognizes include GalCer, 3′-sulfogalactosyl-Cer, GM3, and Gb3 globoside [349]. Gb3 and Gb3 analogues are thus HIV inhibitors [350].

GSL recognition to the HIV-1 gp120 has been a main issue for a long time with regard to the HIV infection to non-lymphoid cells, $CD4^+$ T cells and monocytes. The P^k blood group antigen (or globotriaosylceramide Gb3) has been evaluated as an inhibitor against infection of HIV-1. Seemingly, the Gb3 synthase or α-galactosyltransferase (*A4GALT*) is an indicator for the HIV-1 resistant phenotype of the host cells. HIV susceptibility is linked with the natural devoid of Gb3 expression because individuals with the $P_1{}^k$ phenotype expressing Gb3 are resistant to HIV infection [351]. This is probably due to Gb3 capacity to interact with and compete for the co-receptor-binding region of chemokine present on the V3 loop of gp120. This status prevents gp120 interaction with the co-receptor of chemokines and consequently inhibiting HIV-host cell fusion [352]. Rather, other series GSLs of GM3 and GalCer were regarded as helpers of HIV infection. Gb3 is not expressed in human T cells, but HIV infection induces T cell Gb3 synthesis [283]. Gb3 has several GSL receptor ligand interactions in the interaction of Gb3 as a receptor for the E. coli verotoxin (VT) [353]. For the HIV infection, several cofactors facilitate gp120 binding to cells. Membrane GSLs including GalCer constitute one well-known group of cofactors [354] and this information is helpful in the combinatorial design of therapeutic GSLs derivatives lipid raft targets. Moreover, binding mechanisms of pathogenic viral infection in cells can be easily explained.

V3 loop domain in the HIV gp120 glycoprotein basically interacts with the host GSLs. The V3-loop region of membrane glycoprotein gp120 in HIV-1 is suggested as a ligand for viral attachment, although its precise recognition to target cells is unclear. In the experiment that T cell and macrophage tropically infectious HIV-1 such as X4 and R5 strains, respectively, were used, V3 peptide binds to host cell PM GSLs and exhibited to inhibit the infection even in the absence condition of gp120-CD4 recognition [349]. When the synthetic peptides mimicking the sequence corresponding to the 15–21 amino acids in the V3-loop domain in HIV-1 X4 strain and HIV-1 R5 strain were prepared, the peptides competed and consequently inhibited the infectious entry of the two HIV isolates. The surface GSL-binding HIV V3 peptides potentiate the HIV-1 entry and can be used as a target for the HIV viral entry blocking. It has previously been evidenced that the GSL-binding motif of gp120 carries the amino acid motif of XXXGPGRAFXXX [355]. Interestingly, such similar sequences are also found in some other soluble proteins like synucleins and galectins [356] as well as transmembrane receptor proteins mentioned above. The gp120-type recognizable domain is also found on the extracellular region of the TNF-α receptor super-family with a hairpin structure that contains two aromatic residues (F133 and F134) exposed to the solvent for PCI [357]. The two aromatic amino acid residues directly bind to Gb3 and LacCer, while lowly interact with Gb4 and GD3, too. Therefore, gp120-type GSL-recognizing domains are conserved, as the amino acid region is nested in two α-helices. In addition, the key amino acid Phe residue functions to make docked sugar residue of the GSL glycan with phenylalanine. The carbohydrate-aromatic amino acid residue interaction is noncovalent and observed at axial CH groups on the carbohydrate residue's cyclic structure. If the aromatic amino acid residues are substituted with alanine residue, their binding capacity is greatly reduced [251]. The best glycan residue, giving the high binding affinity to the target protein at aromatic amino acids, is specified to the second Gal residue on the Cer-Glcβ1–4Gal sequence in LacCer [358]. Nonetheless, the GSL-recognizing gp120 V3 loop preferentially binds to Gb3 rather than GM3, although both Gb3 and GM3 are synthesized by the common structure of "Cer-Glcβ1–4Gal"-core, indicating the specific glycan linkage specificity (Fig. 4.2). If GSLs lack for carbohydrate head group like just SM, its binding to the protein region is defective. As a similar case, the extracellular domain of serotonin-1a receptor carries the LNKWTLGQVTC amino acids sequence, which is the well conserved sequence in the serotonin receptor family. The motif amino acid sequence carries the basic amino acid at Lys101, aromatic amino acid at Trp102, and turning amino acid residues at Gly105 in the GSL-binding region.

Fig. 4.2 Ganglioside GM1

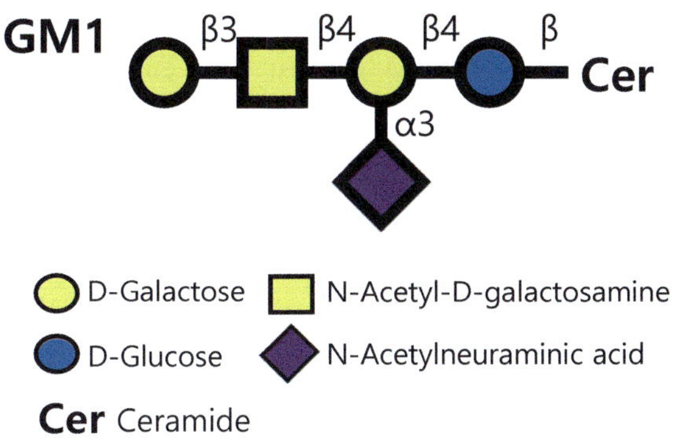

4.2 GM1/GD1b/GT1b-Polyoma Virus Infection

Due to the characteristic linkage diversity of carbohydrates, virus entry and infection use such glycans. This is carbohydrate-specific nature with the broad structural motifs, now seen in proteins and nucleic acids. In addition, although polyoma virus and JC virus recognize SAs, each virus exhibits each distinct specificity. The *Polyomaviridae* for polyomaviruses (PyV) is a member of icosahedral, nonenveloped, dsDNA viruses that cover human BK polyomavirus (BK-PyV), JC polyomavirus (JC-PyV), polyomavirus of Merkel cells (MC-PyV), murine polyomavirus (mPyV), and SV40 [359]. The capsids are composed of 72 pentamers to form the icosahedral capsid structure that is composed of VP1, VP2, and VPs proteins. Polyomavirus capsid protein VP1 is a pentameric protein consisting of 360 proteins. Thus, the main capsid protein is the VP1. Each VP1 pentamer recognizes a VP2 or VP3 in the capsid interior and encases dsDNA genome. Upon virion interaction with a receptor in the cell surfaces, the virions are subjected to internalization and transportation to the cellular organelle, ER. For example, SV40 or BK-PyV entry into cells is quite similar together and they use caveolae-mediated endocytosis to the ER for uncoating. Polyomaviruses recognize sialylglycans on the cell surfaces. BK-PyV as an opportunistic pathogen isolated in 1971 causes severe immunosuppression. Immunosuppressive patients receiving organ or bone marrow transplantation exhibit lytic propagation like polyomavirus-associated nephropathy and hemorrhages. BK-PyV recognition to cellular receptors is not clearly understood. The current knowledge of polyomavirus is based mainly on Rhesus monkey-infectious SV40 and mPyV. Polyomavirus penton contains a minor VP-2 protein copy, which is a capsid protein and a main VP-1 capsid protein with 360 copy numbers or VP3 variant truncated at the N-terminal region of VP2. Polyomavirus capsid pentons bind to surfaced sialylglycans on the host cells. The main BK-PyV penton receptors are GD1b and GT1b. However, BKV VLPs can bind to GAGs on target cells [360] in a manner of GAG-capsomer recognition.

Because BKV genome lacks any polymerase genes, BK-PyV DNA replication exclusively relies on the host system for virus replication. Therefore, BK-PyV intracellularly traffics the ER region through host factors. GD1b or GT1b-bound BK-PyV on the PM of host cells enters host cells through a caveolin-driven endocytosis pathway. The endocytosed BK-PyV enters the endosomal organelle. Apart from BKV, others such as SV40, mPyV, and CTx toxin are also known to penetrate to the cell through receptor-caveolin-endocytosis. Cytosolic BK-PyV is further delivered to nucleus region by the minor capsid proteins through the α/β pathway. VP1 protein pentamers interact directly with the carbohydrate part of gangliosides, GSLs with SAs [361]. SV40-expressed VP1 protein preferentially recognizes the GM1, which is a binding receptor for viral attachment and entry, while mPyV strain requires other ganglioside receptors of GD1a and GT1b as binding receptors, and BK-PyV recognizes GD1b and GT1b as receptors. Interesting, affinity for the GM1 ligand-recognition site as a single site is relatively at a low level (K_d 5 mM). NeuGc and NeuAc are able to serve as receptors of SV40 [362]. The SV40 VP1 showed higher affinity for NeuGc-GM1 than NeuAc form and this indicates SV40's preference to non-human cells such as monkey cells rather than NeuAc-expressing human cells. JCV can bind to a pentasaccharide of glycolipids and glycoproteins as an attachment and infection receptor [363]. The pentasaccharide usage, but not gangliosides, indicates the different preference of viruses as in the entry way of JCV and restriction of cell tropism for infection. Also, it was reported that JCV receptors utilize GM1 and GD1b as the receptors, but also the α2,6-linked Lac-series tetrasaccharide C receptor [364]. JC-PyV utilizes either an SAα2,3 or SAα2,6-attached glycans to interact with glial cells for infection [365] and a GT1b is also potentially involved [366]. But, BK-PyV utilizes the SAα2,3-linked glycans only to enter and infect cells. The GD1b and GT1b are also functioned for BK-PyV infection to kidney cells [367].

The binding pocket domains on VP1 of all the mouse polyomavirus of MC-PyV and mPyV, B-lymphotropic polyomavirus (LpyV) as well as polyomavirus 9 of human (HPyV9) bind to SAs [368–370]. The mPyV recognizes α2,3-linked SA. These sialylglycans were later demonstrated to present in the GD1a and GT1b carbohydrates as the glycan receptors for mPyV infection. In contrast, if the distinct region on the extended VP1 surface loops has the broken Neu5Ac-binding site, human polyomaviruses HPyV6 and HPyV7 bind to non-sialylglycan receptors [369]. Among JC-PyV strains, WT3 VLP strains are known to recognize Gt1b, GD2, GD1b, GD1a, GM2, GM1, and asialo-GM1, while Mad-1 VLP strains recognize the related forms of GM3/GD1a/GD1b/GD2/GD3/GT1b/GQ1b gangliosides; however other GM1/GM2 are not recognized [364]. Despite the research information of JC-PyV ganglioside recognition, the relevant recognition of JC-PyV to carbohydrate structures in the JC-PyV infection is still unknown in detail.

mPyV binds to host cell surface gangliosides and the α4-integrin receptor via the VP1 capsid protein. Receptor binding by mPyV suggested that mPyV binds to α2,3-linked sialylated oligosaccharides present in the GD1a and GT1b as the biologic receptors for mPyV. As an increasing knowledge on viral recognition proteins with sialylcarbohydrates is available in recent years, ganglioside and integrin are the receptor for mPyV infection and required for activation of the PI3K pathway [370]. However, how do gangliosides and α4-integrin play their roles in mPyV infection is not answered. Gangliosides are needed for PyV trafficking to the ER [371]. Human *Trichodysplasia spinulosa*-associated polyomavirus (TSPyV) known as a causing agent of a skin disease in immunocompromised individuals interacts for attachment and infection with glycolipids, but not N- and O-linked glycoproteins [372, 373]. Whether TSPyV directly causes a human pathological infection through molecular binding, recognition, attachment, viral entry, penetration, and intracellular transportation in TSPyV infection is unclear. Using the high-resolution technology of X-ray crystal structures on the VP1 capsid protein complexed with the GM1 glycan, α2,3-sialyllactose (SL) and α2,6-SL, the terminal SA has been demonstrated to bind to the TSPyV VP1. This SA binding sites are commonly conserved in other polyomaviruses [373]. Mutation of SA-binding amino acid residues contributes to reduced cell recognition, attachment, and infection, indicating TSPyV VP1-glycan interaction as a PCI type. GM1 is easily converted by enzymatic desialylation of gangliosides of GD1a/b and GT1b by a neuraminidase as a molecular switch for local function. This may be linked with the GM1 utilization as such viral receptors.

4.3 GM2-GD1a-GT1b-GQ1b-Neolacto-Series GSLs-Paramyxoviruses (Newcastle Disease Virus, Respirovirus, Mumps Rubulavirus, and Avulavirus)

Paramyxoviridae belongs to a family of the order Mononegavirales. This family includes Aquaparamyxovirus, Avulavirus, Ferlavirus, Henipavirus, Morbillivirus, Respirovirus, and Rubulavirus. Paramyxoviruses are nonsegmented and enveloped viruses as the negative-strand RNA viruses. The viruses attach and enter the host cell through surface receptor by glycoprotein (G), hemagglutinin (H), or hemagglutinin-neuraminidase (HN). The attachment and recognition proteins of G, H, HN, and fusion protein (F) are used as the target antigens of neutralized Abs. The MuV consists of genomic 15,384 bp nucleotides. SA-containing glycolipids are also involved in the first stages of the paramyxovirus life cycle and a key viral biomarker of host range. The paramyxovirus belongs to the enveloped viruses and enters to the host cells via specific recognition of the host cells and consequent PM fusion. The viral entry and recognition protein of certain paramyxoviruses including Avulavirus, Respirovirus, and Rubulavirus utilize the specific binding to SA residues on cell surfaces. The paramyxovirus receptors are mainly elucidated in molecular and chemical levels from the Sendai virus (murine parainfluenza).

SA-bearing glycans are involved in the initial entry step of life cycle of the enveloped paramyxovirus. The attachment and entry to the target cells commence with their recognition to the host cell surfaces and followed by PM fusion. Virus recognizes the surface receptors by its attachment proteins resided on paramyxoviruses. HN glycoprotein recognizes cellular SA molecules and exerts sialidase activity and fusion activity. For example, the Sendai virus attachment and recognition proteins are called hemagglutinin-neuraminidase (sialidase), HN glycoprotein, which recognizes and binds to host cell PM, although the virus-induced membrane fusion is still not well understood yet. For example, Sendai virus uses SAα2,3-linked gangliotetraosides such as neolacto-series gangliosides, GQ1b, GT1b, and GD1a as the attachment receptors, but not SA-lacking GM1 or GD1b on the terminal Gal residue. However, limited information is available for the entry receptors for currently reported other paramyxoviruses, compared to those of Sendai virus. However, in the human parainfluenza virus, type 1 virus or type 3 virus binds to N-acetyllactosamine-bearing glycans terminated with SAα2–3Gal.

For a model of other paramyxoviruses, GM3 is a Newcastle disease virus (NDV) receptor but not for Sendai virus receptor [332]. Sendai virus recognizes gangliosides, which terminally contain the NeuAc residues attached to Gal residue. As the Sendai virus receptor, GD1a influences lipid polymorphism and isotropic structure formation in monomethyldioleoyl phosphatidyl ethanolamine. Therefore, Sendai virus fusion with membrane liposomes is ganglioside dependent [333]. Apart from Sendai virus, although the receptors for other paramyxoviruses remain unknown, some sialylated glycoproteins, not gangliosides, are used as specific paramyxovirus receptors. For example, human parainfluenza viruses recognize Neu5Acα2–3Gal-attached N-acetyllactosamino-glycans. For specifically, the human parainfluenza viruses of type 1 of hPIV1 as well as type 3 of hPIV3 have been specified in the carbohydrate recognition, where they preferentially recognize LacNAc-glycans with terminally attached NeuAcα2–3Gal. hPIV3 binds to NeuAcα2–6Gal- or NeuGcα2–3Gal-carrying glycans [331]. As a quite similar type of viruses, Orthomyxoviruses, which are influenza viruses, also recognize SA-carrying glycans, although each influenza strain needs each specific SA linkage (α2–3 or α2–6) or SA type (NeuAc, NeuGc, or 9-O-Ac-NeuAc) [330]. Paramyxovirus NDV is not restricted to only gangliosides for binding to the host cells, but potentially binding to SA-containing glycoconjugates and sialoglycoproteins of host cell PM [374].

For NDV case which is relatively well established to date, two transmembrane glycoproteins, HN and fusion (F) protein [323], are expressed in viral envelope. The NDV uses the NeuAc form. HN recognizes cell surface SA-attached receptors and F protein facilitates the viral envelope fusion into the host cell PM [324]. The binding of HN to the host cell receptors induces changes in conformational structures of the protein HN and consequently induces the fusion capacity of F protein. In experimental binding assay between NDV and glycolipids, several GSLs including GM3, GM2, GM1, GT1b, GD1b, and GD1a are bound to the NDV virus [335]. In molecular level, NDV recognizes the terminally attached SA residue in GM3, GD1a and GT1b as well as the internally linked SA residue in GM2, GM1, or GD1b. However, GQ1b is not bound to NDV, although GQ1b occupies both terminal and internal di-SA residues. In contrast, GQ1b is known to bind to Sendai virus [334]. Apart from the ganglioside recognition, NDV binds to sulfatides and asialo-GM1, but not to Gal-Cer or Lac-Cer.

Mumps rubulavirus or Mumps virus (MuV) is a deafness, encephalitis, meningitis, and parotitis-causing human pathogen. MuV is the genus Rubulavirus of a family of Paramyxoviridae. Mumps vaccination is combined with measles-mumps-rubella (MMR) vaccine at ages of 12–15 months and measles-mumps-rubella-varicella

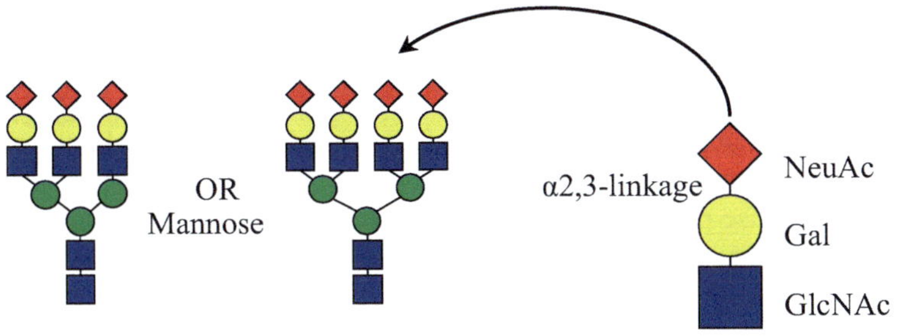

Fig. 4.3 MuV HN binds to α2,3-SA linkages and prefers a SAα2,3-linked trisaccharide as a receptor. Additionally branched α2,3-SA-containing glycans are not strong to bind to MuV HN

(MMRV) vaccine at the ages of 4–6 years. MuV is an aerosol-transmitted human pathogen, belonging to Paramyxoviruses as enveloped and non-segmented negative-strand RNA viruses. The MMRV includes four genotypes (A, B, H, and N). Viral proteins include hemagglutinin-neuraminidase (HN), hemagglutinin (H), glycoprotein (G), and fusion protein (F). Each genotype conveys conserved α-helix domain in C-terminal receptor-binding region. MuV HN binds to α2,3-SA linkages and prefers a SAα2,3-linked trisaccharide as a receptor for mumps virus. The branched α2,3-SA oligosaccharides are weak in binding to viral HN, when compared to α2,3-SA trisaccharide (Fig. 4.3). Among 3SL and 6SL, 3SL prefers to bind to MuV-HN domain. Aromatic amino acid residue is crucial for 3SL recognition. Only α2,3-SA glycans are the MuV-HN receptors. N-/O-glycoproteins and GSLs are utilized as receptors for MuV HN.

MuV tropism specifies glandular tissues and CNS through carbohydrate receptors. MuV hemagglutinin-neuraminidase (MuV-HN) protein recognizes sialyl LewisX (SLeX) and the GM2 carbohydrate structure. SLeX and GM2 carbohydrates share the common trisaccharides of 3'SL. Thus, the binding site of MuV-HN recognizes both SLeX and GM2. GM2, SLeX, and 3'-SL sugar block MuV entry to host cells [375]. The α2,3-SA glycans such as SLeX and SA-LacNAc are abundantly present in most tissues and GM2 is present mainly in neuronal cells, tissues, or adrenal glands, indicating MuV tropism. The diverse MuV recognition of the different carbohydrate receptors indicates its tropism and pathogenesis. MuV-HN favors SAα2,3-linked glycans over SAα2,6-linked glycans. The MuV-HN binds to both sugar types of GSLs and glycoproteins as receptors. Amino acid residue at Tyr369 position of MuV-HN recognizes 3'-SL and the SAα2,3-linked trisaccharides in all MuV species of the genotypes A, B, G, and L.

The specifically designed SA mimics can inhibit the HN or receptor-recognizing capacity of viral proteins can be designed as a future antiviral therapeutic strategy. For example, the precise information of paramyxovirus entry receptors and the viral entry to the cell surface will create a new vision.

4.4 GT1b/GQ1b-Rabies Virus (RABV) Interaction

Virus infection of cells is carried out as a stepwise process with initial binding to receptors in host cell surfaces, PM fusion of enveloped viruses or PM internalization and PM penetration, trafficking and replication, progeny production of virion, and extracellular release of viral particle into the cytoplasmic spaces. Therefore, receptors determine viral pathophysiology and tropism with tissue specificity. Rabies virus (RABV) is an infectious pathogen to neurons in the nervous system of mammals. RABV is an enveloped and negative sense ssRNA virus. This virus belongs to the Lyssavirus genus of Rhabdoviridae family. This Rabies-causing agent raises 100% fatality, indicating neurotropic RABV threat. A variety of candidate receptors are known for carbohydrates, gangliosides, NCAM, nicotinic acetylcholine receptor (nAChR), metabotropic glutamate receptor subtype 2 (mGluR2), and p75 neurotrophin receptor (p75NTR). After attachment to host cells, RABV is endocytosed via clathrin-mediation and transported through retrograde trafficking endosomal vesicles. The RABV consists of five viral proteins such as matrix protein (M), nucleoprotein (N), phosphoprotein (P), RNA polymerase (L), and glycoprotein (G). The linear RNA encapsulated by N proteins yields the nucleocapsid complex. The nucleocapsid-RNA complex with the P and L proteins forms a ribonucleoprotein (RNP). The viral coat membrane protein M wraps the RNP and protein G-based spikes bind to receptors and fused with PM with the protein M, potentiating entry of virus into host cells.

RABV is transmitted via animal bites. Virus particles enter the nervous system through a sensory nerve, the neuromuscular junction, and motor axons of the muscle surface. The viral envelope contains two G and M proteins. RABVG has a form of a trimeric type I glycoprotein structure with a 505 amino acid. The G trimer of RABV recognizes cells. RABVG mediates the virus transport to the CNS via the retrograde pathway. nAChR has been known as a receptor for muscle infection. RABV then uses nAChR, NCAM, p75 neurotrophin receptor named p75NTR, and sialyl GSLs for infection to host cells. Muscular nAChR binds to RABVG for RABV infection. Rabies virus adhered and attaches to the PM as the first step. For the cellular components in the attachment, gangliosides have also been known in rabies virus infection to chick embryo cells. Upon treatment with neuraminidase, the cells were desialylated without viral infection. However, cells treatment with gangliosides restored ganglioside incorporation and access to rabies virus infection. Among gangliosides, GT1b and GQ1b have been reported as the best receptor candidates and GD1b was the moderate candidate, whereas GM1 or GM3 was the poor one [337]. In the neuron membranes, polysialogangliosides are the receptor of rabies virus infection. In the healthy foxes and rabies virus-infected foxes of brain, glycoproteins and polysialylated gangliosides have been reported to be related [338].

4.5 GD1a-Adenovirus Infection for Keratoconjunctivitis

Adenoviruses are human pathogens. Human adenoviruses are problematic with immunocompromised patients in healthy individuals at the eye, gastrointestine, and respiratory system. Adenoviruses have seven species of A to G group with 57 to 90 types. The severe infectious species include adenovirus AdD56, AdB7d, and AdE4 through epidemics of infection. Receptor tropisms of adenoviruses are based on coxsackie and adenovirus receptor (CAR), CD46, desmoglein 2, or SA-containing carbohydrates [376]. Adenoviruses D species is mostly common with 35 of 57 canonical types. The species D causes epidemic keratoconjunctivitis (EKC) disorders. The EKC adenoviruses include Ad-D8, Ad-D37, and Ad-D64 in humans. Adenovirus is the dsDNA and feasible for genetic modification and therefore, this species is attractive for genetic applications to regress cancers through genesis of oncolytic viruses and vaccination vectors. Virus attachment, recognition, binding, and interaction with the cell surface indicate determinant property for tropism. Understanding of the precise recognition and interaction opens a new and potential therapeutic strategy to target for antiviral therapy. Attachment factors on the surfaces of host cells contribute to viral concentration. Certain viral receptor or co-receptors facilitate virus interaction with the host cell surfaces, recognition, binding, entry, and infection to host cells. Adenoviruses such as CAV-2, Turkey adenovirus 3, and human Ad-G52 interact with SA residue through the fiber knob lateral region. Human Ad-D8, Ad-D19p, Ad-D37, and Ad-D64K viral fiber knob proteins also recognize SA through their apical regions.

A residues as Neu5Acβ2,3-D-Galpβ1,3-D-GalNAc1,4-(Neu5Acα2,3)-D-Galpβ1,4β-D-Glcp1,1-Cer. Previously, GD1a was reported to modulate the cytokine, GM-CSF, or leukocyte growth factor function and it has been known to modulate GM-CSF-induced cell proliferation [377]. GD1a regulates the expression level of the GM-CSF receptor (GMR). GMR is a heterodimeric protein and consists of a ligand-binding region for GMRα chain and a constant β chain (βc). GMRα is composed of two distinct isoforms and is polymorphic. Two distinct isoforms are known: (1) transmembrane dimerizes with βc and leads to signaling and (2) soluble variants lack the transmembrane domains. GMRα isoforms are synthesized by C/EBPα transcription factor upregulation by GD1a. GD1a colocalizes with GMR in monocytes. Exogenous GM3 or GD1a stimulates GM-CSF-activated cell growth, as monocytic cells synthesize predominantly GD1a and GM1. Reduced ganglioside synthesis decreases in GM-CSF-stimulated proliferation. The carbohydrate part of GD1a is used as a receptor responsible for adenovirus infection of adenovirus type 37 (Ad37) causing EKC disorders [378]. EKC is a highly contagious ocular disorder raised by three D adenovirus species. General pathological symptoms of EKC involve edema, pain, lacrimation, and vision regression. Other types of adenoviruses interact with host cells through binding to CD46 or the CAR named Ad37 protein, which recognizes SA-linked molecules. Using glycan microarray technology, the receptor-binding knob, which is present in the Ad37 fiber protein, has been found to bind a branch type of hexasaccharides in the GD1a having two terminal SAs. GD1a or GD1a-specific Mabs block virus Ad37 binding capacity and infection to eye corneal cells. GD1a carbohydrate structure, but not the ganglioside GD1a itself, was organized with glycoproteins. Many D adenovirus species such as the EKC bind to the GD1a glycan–contacting residues. Heterogeneity in GD1a-depending infectiousness of several related adenovirus types is shown, although the adenoviruses, which cause the EKC, utilize the carbohydrate part in GD1a as an infectious receptor on the cell surfaces, further defined conclusions seem to be needed to finely define the function of GD1a as a receptor for non-EKC-causing D virus species or the relatedness. Using other experimental technologies including molecular imaging, modeling, NMR, and X-ray crystallography diffraction, the two terminal SAs were shown to dock into three SA recognition sites on the trimer molecule of Ad37 knob. The knob-GD1a glycan binding shows a high affinity as confirmed by surface plasmon resonance, giving a solution toward the future development of SA-based antiviral agents against EKC. The discovered GD1a carbohydrate glycan as an infection receptor for Ad37 may give a clue for inhibitory drugs. Drug compounds can be remodeled on SAs likely to Tamiflu in influenza A viruses, as highly useful antiviral therapy of influenza [377, 379]. The PCI between Ad37 knob and GD1a resembles with the influenza hemagglutinin (HA)-sialylglycan motif interaction [363]. For example, the SA-recognizing region is present on the influenza HA tip, but the SA-recognizing site on the knob is present around the central cavity.

4.6 GalCer-Adeno-Associated Virus (AAV)-Bovine AAV (BAAV) Interaction

Adeno-associated virus (AAV) belongs to the genus Dependovirus that uses a helper virus for replication. AAV is the ssDNA virus of family Parvoviridae, which functions as a replication helper for either herpesvirus or adenovirus. They are not toxic and are not related to any disease. AAV capsids consist of 60 protein subunits with VP1, VP2, and VP3 proteins, as alternatively spliced forms, which are the same gene regions. AAV utilizes diverse structures of cellular carbohydrate for attachment and infection. Type 2 AAV (AAV2) binds cell surface heparan sulfate proteoglycans. Bovine AAV (BAAV) also utilizes PM gangliosides for attachment and infection. AAVs are reported to recognize diverse glycan structures of proteoglycan heparan sulfate, N-type and O-type sialylglycans, and GSLs expressed on cell surfaces for recognition, attachment, binding, entry, penetration, and infection, but also for the virus life cycle. BAAV requires gangliosides and β1,4-GlcNAc residues on glycoprotein gp96 for virus attachment [325, 380]. Apart from the infection, BAAV uses carbohydrate interactions for infection or transcytosis, because it can also pass through epithelial barriers and endothelial barriers using cell surface carbohydrates-utilizing transcytosis. Transcytosis has been known for macromolecules and pathogens, where transcytosis event is similar to the macromolecular movements from one side of a cell to the other sides [381]. In HIV and *S. pneumoniae* transcytosis, surfaced Gal-Cer and poly(IgA) are receptors, respectively [380]. The carbohydrate interaction for BAAV transduction movement or transcytosis movement requires a trimeric β1,4GlcNAc residue. A β1,4GlcNAc containing membrane glycoprotein gp96 is a receptor for BAAV transcytosis.

4.7 GM3/GM1a/Histo-Blood Group-Rotavirus Infection

Rotavirus belongs to the Reoviridae family as dsRNA viruses and causes animal and human infantile gastroenteritis. Small intestinal epithelial cells are targeted by the viral infection. The rotaviral particle or virion is composed of triple layers with a core segmentation genome, protein shell in internal region, and 7-trimeric virus protein (VP) capsid in which VP4 spike is penetrated. The serotype-dependent VP7 and protease-sensitive VP4 determine the rotavirus classification to type P rotavirus (VP4) and type G rotavirus (VP7). The VP4 protein of type P virus is subjected to the proteolytic cleavage to generate the N-terminal VP8 protein and C-terminal VP5 protein. VP8 attached to the dimeric VP4 spike head recognizes surface glycans of host cells at the early stage of life cycle. VP5 recognizes the α2β1 integrin to elicit infectious event and VP7 recognizes the integrins of αxβ2 and αvβ3. The α4β1 and α4β7 integrins are crucial for rotavirus entry to some host cells. Integrins-glycan receptors clustering in lipid rafts microdomains with HSP70 also are involved in the entry of rotavirus to host cells. The rotavirus receptors are also topical trend of rotavirus research, because entry of rotaviruses to host cells is called complexed process. The host receptors determine virus tropism to infection. Several events during evolution, including gene rearrangement, point mutation, rotavirus genome segmentation, and wide type G and P range potentiate gene reassortment, contributing to increase in genetic diversity. The most variable region of structural proteins is the VP8, allowing alteration in glycan receptor usage by rotaviruses as rotavirus entry mechanism.

The entry process involves sequentially interacted pathways with cell surface gangliosides with one or more SA residues. GSLs are a heterogeneous family of amphipathic lipids found on mammalian PMs. Rotaviruses are etiologically causing factors of gastroenteritis in animals and humans through the entry and infection to intestinal epithelial cells [382]. The *Reoviridae* family is dsRNA viruses and the rotaviral particle is layered in triple sets and made of an inner shell protein and an outer SP-4 with spike capsid VP 7 trimers. Thus, the outer layer contains VP4 spike proteins projecting through an inner VP7 shell protein. From the serotype-based VP7 glycoprotein and *protease*-susceptible VP4, rotaviruses can be further subclassified to G type virus for VP7 and P type for VP4 virus. VP4 is associated with rotavirus-cell attachment, while VP4 and VP9 determine serotype specificities of rotavirus P and G serotypes, respectively [383]. VP8 located on the dimeric VP4 head spike binds to carbohydrates on cell surfaces. VP5 is conformationally shifted affordable for host cell membrane penetration [384]. VP4 cleavage by proteolytic degradation is important for conformational shifts for virus entry. VP5-carrying rotaviruses thus bind to the α2β1 integrins and potentiate infection to cells [385]. VP7 enters rotavirus cells via the αxβ2 and αvβ3 integrins, while the α4β1/7 integrins in some rotavirus can be used for specific host cell types [386].

Rotavirus entry is initiated by attachment to a glycan receptor of a glycolipid head group with subsequent vesicle formation. Penetration is operated by the uptake vesicle. Gangliosides are involved in successful endocytosis at the PM, and the mobile movement of the GSLs in the cell PM may further potentiate rotavirus entry. The rotavirus infection glycan receptors are specially interesting in recognition, attachment, endocytosis, host tropism, and pathogenic disease. The SA binding is decided by the VP8 amino acids at 187 and at 157 as a Sia binding site. Rotavirus also uses the histo-blood group antigens (HBGA) for its recognition receptor molecules. The HBGA is indeed asialo-glycans appeared on epithelial cells and erythrocytes, and also in certain mucosal secretions [387]. The diversity of the host receptor glycans is recognized by rotaviruses, and the GM1a and HBGA are the major receptors. Then, the question how the ligand spike protein of rotavirus interacts with the GM1a receptor at the molecular structure and mechanistic levels is interesting in understanding innate intestinal immunity and provides rationale to treat and prevent the rotavirus infection. The VP8 of human rotavirus P subtype [388] recognizes Le^b and H-type I HBGA, whereas VP8 of P6 subtype of neonatal ST-3 strain binds to only the H type I HBGA but not others [389]. Rotaviruses enter and penetrate host cells by disrupting membrane structure. RRV VP8 binds to GM1a and GM3 as a crucial step for binding of VP8 of the known many rotavirus strains.

Their host cell receptor and attachment ligand are glycans and the virion spike protein, respectively, toward precise host infection by rotaviruses. Rotaviruses prefer ganglioside to enter the host cells. Rotavirus receptors utilize sialic acids, gangliosides, and HBGA and the receptor-glycan binding induces VP8 protein conformational change. VP8 mediates virus entry to host receptors via the co-localized integrins-glycan receptors complex with additional membranous lipid rafts and cognate HSP70 [390]. NeuAC and NeuGc and α2β1 integrin are known as rotavirus cellular receptors by recognition with virion spike protein, VP-4 [391]. Rotavirus's HBGA engagement can be influenced by commensals, fucosylation, and immune system. Engagement of both gangliosides and HBGA receptors can further potentiate their adaption to emerging new hosts. SA residues present in terminal and internal sites in gangliosides and HGBA can be easily bound by VP8 in rotavirus host cell invasion. Human rotaviruses commonly engage GM1a as their receptor while outstanding variations identified in human-infectable rotaviruses determine their capacities to utilize A-type of HBGA, indicating the specific tropism. Therefore, viral host tropism derived from each distinct glycan receptor usage will direct the applied research. The plastic variation in VP8 conformational structure and preference to glycan receptors will clarify the rotavirus adaption acquisition to newly accessible hosts, regarding non-human-to-human transmission via gene reassortment events.

Human rotaviruses utilize GM1a ganglioside as a receptor, although GM1a engagement is not restricted to the human-infectable rotaviruses and there is also not a species specificity in this rotavirus type. Some rotaviruses in humans and UK rotaviruses in bovine utilize GM1a form of GSLs during host cell infection and invasion by human rotaviruses. Therefore, GM1 tropism of human rotaviruses is not certain in its mechanism. The only assumption is that terminal SA residue and internal (or branched) SA residues may be the determinants for rotavirus species. The infection level of animal rotaviruses in host cells is decreased when neuraminidases are treated. The rotavirus VP8 binds to SA residue. The amino acid position 187 of VP8 determines SA-binding preference and the amino acid position at 157 determines SA-binding affinity. Animal-infectable rotaviruses use two different SA types of Neu5Gc and Neu5Ac as their binding carbohydrates. However, RRV known as a simian strain prefers Neu5Ac type, while swine type CRW-8 strain and bovine type NCDV strain prefer Neu5Gc type as a recognition SA type. Rotavirus VP4 binds α2,3-SA linked to the GM1 glycan chain, as the Wa rotavirus VP4 recognizes the internally located NeuAc on GM1 oligosaccharide in an independent mode of α2β1 and internal SA recognition. Another cholera toxin B (CTxB) known as a specific GM1-binding ligand diminishes rotavirus Wa infection to host cells, because the GM1 is the receptor both of CTB and VP4. GM3 gangliosides are also suggested to be appropriate receptors for the porcine rotavirus OSU strain [392, 393].

On the other hand, among the rotaviruses, the two types of sialidase-sensitive, simian SA11 and bovine NCDV as well as sialidase-insensitive, bovine rotavirus strains are known. The infection by the simian rotavirus SA11 strain is mediated by the cell surface SAs species. GM1 ganglioside blocks the rotavirus SA11 infection to the host cells [394, 395]. The known three subtypes including SA11, NCDV, and UK rotavirus strains recognize non-acidic GSLs of gangliotetraosyl-Cer (GA1, another name of asialo-GM1) and gangliotriaosyl-Cer (GA2, another name of asialo-GM2), giving the distinct binding specificities. The viruses also bind to sialylneolactotetraosyl-Cer and GM2 and GD1a. Among them, strain UK rotavirus only can bind to NeuAc-GM3 and GM1. Sialyl-Gal (NeuGc/NeuAca2–3-Galbeta) is the binding site. Instead, only SA11 and NCDV rotavirus strains can bind to NeuGc-GM3. Thus, sialidase-susceptible strains bind to externally located SA residues in gangliosides, while sialidase-insensitive strains bind to internally located SAs of gangliosides [396, 397].

4.8 GM1/GM2-Reovirus Infection of Reoviridae Family

Several viruses primarily use glycans as a receptor, while some viruses like reovirus use glycans as an initial attachment site prior to recognition of a protein receptor to strengthen the binding level. Reoviruses bind to surface carbohydrate receptors as the first infectious step. Mammalian reoviruses sero-typically disseminate with the host by specific glycan recognition. Each reovirus serotype recognizes each specific glycan but the molecular function of each glycan is not well understood in pathogenesis. Reoviruses bind to surface carbohydrate receptors as the first infectious

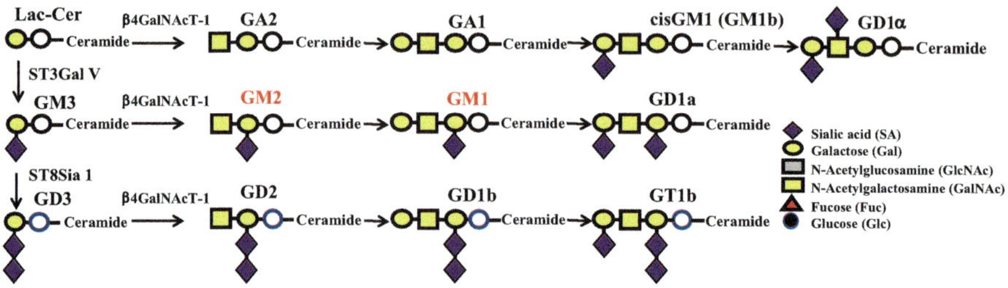

Fig. 4.4 Structures and synthetic pathways of GM2 and GM1 as well as the related gangliosides

step. Each reovirus serotype recognizes each specific glycan but the molecular function of each glycan is not well understood in pathogenesis. In mammals, reoviruses infect in a serotype-dependent manner in the murine CNS. Mammalian orthoreoviruses (reoviruses) display broad cell and tissue tropism in vivo and also are well-designed to study virus-receptor recognition and interactions. Reoviruses are double-stranded RNA (dsRNA) viruses with ten segments encapsidated in the protein shells of the two types, outer capsid and core capsid, infectious for the mammalian gastrointestinal and respiratory tracts, rarely causing systemic disease in the newborn [398]. For the reoviral pathology, apoptotic events are developed in cells and infected mice. Therefore, inhibition or perturbation of viral disassembly by endosomal acidification inhibitors or proteolytic enzymes abrogates apoptosis or death signaling. Reoviral disassembly activates NF-κB to induce apoptotic signaling in cells [384]. Reoviruses prefer to infect tumor cells, giving a possibility of being used in cancer treatment [399]. Reovirus serotype 1 (T1) propagates through hematogenous infection with a tropism to specifically target hydrocephalus-causing ependymal cells. However, reovirus serotype 3 (T3) propagates through hematogenous and neuronal tropisms to target lethally encephalitis-causing CNS neurons. However, the molecular mechanism underlying the serotype tropism in neuropathogenesis remains unexplained. Reassortant reovirus strains that are evolved to have gene segments of two independent reoviral strains exhibited that the attachment protein σ1-encoding S1 gene gains serotype-specific CNS infection through different receptor recognition. The three reovirus serotypes are currently known. For example, type 1 Lang reovirus is named T1L virus and type 2 Jones reovirus is called T2J virus as well as type 3 Dearing reovirus named for T3D virus, depending on distribution of σ1 capsid protein. T1L can target to infect ependymal cells, causing hydrocephalus [400], where the σ1 150 kDa protein homotrimers are involved in the virus attachment to target cells.

Reovirus T1 and T3 use the identical receptors as proteins, which include Nogo receptor 1 (NgR1) and junctional adhesion molecule (JAM)-A. However, the reovirus T1 and T3 recognize distinct glycans as carbohydrate receptor. Reovirus T1 recognizes GM2, which has terminally α2,3-Neu5Ac and β1,4-GalNAc residues. The carbohydrate structure in GM2 glycan is present in both glycoproteins and the ganglioside GM2 (Fig. 4.4). Therefore, α2,3sialylglycans are used as the carbohydrate glycan receptors for reovirus T3, but not for other reovirus types of glycan binding in T1 reovirus infection. The interaction between reovirus T1 and GM2 occurs in the physiological condition. In contrast, reovirus T3 recognizes the GM3-containing carbohydrate of the SAα2,3-Gal-Glc linkage, SAα2,6-linkage, and SAα2,8-linkage of sialylglycans. Exogenous treatment with GD3, GM3, and 3SL blocks reovirus T3 adherence and attachment to host cells. Because exogenous GM3 carbohydrate does not block reovirus T1 infection to cells, it is suggested that reovirus binding to glycan receptors depends on serotypes. Thus, the SA-binding reovirus T3 can target neurons more effectively than the SA-unrelated strains [401]. The glycan-binding region of T3D σ1 interacts with α2,8-NeuAc, α2,3-NeuAc, and α2,6-Neu5Ac as its carbohydrate receptors, indicating that carbohydrate-recognizing potentials contribute to each specificity of the viral pathogenesis [401, 402]. Binding is made by a bidentate salt bridge type, connecting Arg202 with the carboxylic acid of Neu5Ac residue. The additional lactose backbone additionally recognizes T3D σ1, utilizing a distinct glycan sequence appeared on the surface PM of the host cells [361]. Reovirus hemagglutination shows each serotype dependency. T1 reoviruses can specifically agglutinate human erythrocytes, but not erythrocytes of bovine. In contrast, T3 reoviruses can agglutinate bovine red blood cells, too [403], indicating the distinct carbohydrate-recognizing regions present on the T1 and T3 reovirus are quite different.

Because human erythrocytes express Neu5Ac and bovine erythrocytes express mainly Neu5Gc type, less amount of Neu5Ac, the hydroxyl group of Neu5Gc, has been suggested to favorably recognize a hydrophobic pocket in the type 1σ1 carbohydrate-recognition region. Using structure-guided mutagenesis, a GM2 receptor non-binding T1 reovirus mutant has been made [395]. The mutant virus does not cause hydrocephalus, compared to the wild-type virus in GM2-deficient mice and in cultured ependymal cells of the brain ventricles. Recently, from the glycan microarray, GM2 has also been found as a carbohydrate receptor for T1L reovirus [398], binding to the T1L σ1 head region. GM3 can also bind to T1L σ1; however, GM2 glycan is preferentially used as a carbohydrate receptor for T1L reovirus probably due to the extra-containing GalNAc residue of GM2. The reovirus σ1 is a trimeric fiber composed of three domains of body, tail, and head. The T1 σ1 head region protein recognizes the Neu5Ac and GalNAc residues of the GM2 carbohydrates. Amino acids of Val354, Ser370, and Gln371 residues interact with GM2, because the S370P or Q371E mutation in reovirus σ1 protein is impaired for agglutination of human erythrocytes and infection to host cells. The S370P and Q371E double mutants created for the reovirus T1L σ1 attachment protein do not recognize GM2 but wild-type T1L σ1 does bind to GM2. In the infection of the wild-type T1L and the S370P-Q371E double mutants, glycan-recognition capacity is likely not to largely influence replication in the infected murine. However, the hydrocephalus level of the T1L infection of wild type is high compared to the S370P-Q371E double mutant, indicating that T1L is a less symptom type of murine of hydrocephalus, which GM2 is not produced. Therefore, GM2 recognition is a key factor of each serotype reovirus symptom and is an essential factor of serotype-dependent infection of reovirus. Branched glycan GM1 has been suggested to be essential from the competition with cholera toxin [404]. Except for the GM2 glycan-recognition region in T1Lσ1, another specific T1L σ1-binding site is the JAM-A, although the D1 region of human JAM-A N-terminal region does not contain any glycosylation site [403]. Thus, the carbohydrate receptor is independently engaged and the reovirus-host cells infection may use multiple steps, postulating that the virus first recognizes GM2 with low affinity and thereafter it recognizes JAM-A with relatively higher affinities than the previous one [405], following integrin-mediated uptake of virus.

As well described for the reovirus, entry strategies displayed by the enveloped viruses need multiple recognition, binding, and interaction between virus and cell surface receptors, accompanying with conformational changes in virus-recognizing proteins. Such enveloped viruses involve a recognition-binding-adhesion-enhancing strategy during initial virus-cell interactions, where show the lower affinities between virus and surface carbohydrates [406]. Then, this event is followed by the increased affinities between virus and secondary receptor, eventually viruses entering via membrane fusion or receptor-mediated endocytosis [407]. With regard to the increased affinities, SA is often used as a coreceptor toward virus attachment and tropism. Binding to SA is the first initial step of viral infection. For the next step, high-affinity binding to a secondary receptor isrequired for the further interaction with host cells, giving a conceptionally such tropism, implies for the attachment-enhancing mechanism for reovirus interaction because sialic acid and head receptor expressions determine the virus-specific infection [408]. Higher expression of AS and head receptor indicates more efficient infection. Currently, because reovirus is under investigation for a vaccination vector as well as oncolytic agent development, reovirus-glycan recognition can be applied to the therapeutic trials of reovirus glycan-binding mechanism. Understanding reovirus binding to glycan may allow some clue to apply for development of therapeutic agents.

4.9 Gb4-Parvovirus Infection

The Parvovirinae has five genera as a subdivision. They are amdovirus, bocavirus, dependovirus, erythrovirus, parvovirus, aleutian mink disease virus, bovine parvovirus, AAV2, human parvovirus B19, and mice minute virus, which are based on genome structure and protein structure. Their capsid open reading frame (cap) encodes two or three (depending on the virus) overlapping structural viral proteins (VP) which assemble the $T = 1$ capsid. The non-pathogenic dependovirus replication depends on co-infection with a helper virus of adenovirus, herpes simplex virus, or papillomavirus. The other genera replicate by themselves without helper virus cooperation. AAV1 recognizes both SAα2–3 and SAα2–6 N-glycans. AAV2, AAV3, and AAV13 recognize heparan sulfate proteoglycan (HSPG). AAV4 and AAV5 recognize SAα2–3 O-glycan and SAα2–3 N-glycan, respectively. Bovine AAV binds to gangliosides and chitotriose of β1,4-GlcNAc for infection. GSL globoside/globotetraosylceramide (Gb4Cer) is an infection receptor for B19 parvovirus of humans. Virus-like particles directly recognize globo-series Gb4Cer.

Human B19 parvovirus type belongs to a small nonenveloped virus as an ssDNA type and belongs to the genus *Erythrovirus*, the subfamily *Parvovirinae,* and the family *Parvoviridae*. As a human pathogen, the B19 type species of human *Parvovirus* is known to cause the childhood disease erythema infectiosum [341, 409]. B19 is also problematic in pregnancy of humans and infection of fetal ages causes fetal death with hematological disorders. The viral capsid is composed of two VP1 and VP2 proteins, forming an icosahedral symmetry with 60 structural protein subunits. The two proteins differ in terms of the additionally repeated N-terminal sequence, 227 amino acids, on VP1. B19 virus has an extremely narrow tissue tropism and infects most of human erythroid progenitor cells. B19 virus replicates in the erythroid lineage BFU-E and CFU-E cells. For the B19 virus receptor and internalization, virus infects erythroid progenitor cells targeting its receptor, the P antigen of blood groups [410]. In fact, B19 parvovirus infects the erythroid progenitors in the BM by recognizing the GSL globoside Gb4. Virus binding to Gb4Cer has been demonstrated on thin-layer chromatogram (TLC) data, confirming that the Gb4 globoside carbohydrate structure is crucial for viral attachment, interaction, and entry to the host cells. Anti-Gb4 antibodies specifically inhibit the B19 virus infection to bone marrow-derived mononuclear cells; exogenous Gb4Cer treatment blocks the viral binding to erythroid cells.

Tropically, Gb4 species is present in many different cell types, but predominantly present in erythroid lineage progenitor cells of humans in the bone marrow. B19 virus hemagglutinates but soluble or Gb4 treatment inhibits the hemagglutination. However, the current issue of B19 virus tropism is that B18 virus pathogenicity and erythroid tropism are directly not related with Gb4 expression. The Gb4 level in host cells is not related to B19 viral recognition with the cells but its level seems to be necessary but not exclusively for replicative infection. This indicates that PM molecules may affect the GB4 binding with other GSLs to B19V.

The fact that the Gb4Cer is not recognized in the phospholipid bilayer membrane indicated that Gb4Cer is not the direct actor like a bona fide receptor to B19. Then, it has been known that B19 binding to the cellular receptor globoside (Gb4Cer or Gb4) indeed leads to structural and conformational shifts and alternation of the capsid proteins. This allows the viral accessibility via the N-terminal region of VP1 (VP1u) to host cells. The Gb4 is called P-antigen due to its synonym and its synthesis is well known in certain carcinomas in the testicular tissues. P antigens is the site for B19 infection in cells, as the tumor cells can be infected by B19 infection [411]. However, co-receptors are required for infection in erythroid progenitor cells. During internalization to the host cells of humans, B19 virus binds to the erythroid progenitor cells through Gb4 interaction or blood group P antigen interaction [35]. Because P antigen is largely expressed in erythro-linage types, surface P antigen binds to the virus. But the process is not sufficient for complete entry and infection. It indicates the existence of some co-factors including $\alpha5\beta1$ integrin, and Ku80 is required for the complete infection of B19 [412]. The Gb4Cer as receptor has been reconfirmed for B19 [413, 414]. The oligosaccharide epitope is functioned as a B19 ligand, and this knowledge regarding the viral binding with Gb4Cer opens a new B19-Gb4Cer interaction as the PCI type for virus infection [415]. B19 virus also binds to membrane-associated Gb4Cer, which is a form of the reconstituted Gb4Cer, as a receptor [416].

4.10 SA/GM3-Influenza a Virus Infection

4.10.1 SA-Specific Influenza a Virus

Influenza A belongs to a negative-sense ssRNA virus as an enveloped type of the Orthomyxoviridae family and causes epidemics in humans. Enveloped type of viruses is featured with a lipid bilayer. The lipid bilayer is easily embedded or merged with the target cell lipid bilayer. This is a first step for the viral entry into host cells. The influenza virus is classified depending on the virus-host infectious similarity. Three different Influenza virus species are classified. A, B, and C virus types are known for the influenza. Human influenza virus A/B types are the causative factors of seasonal influenza emergence in humans. Zoonotic influenza A type virus infects humans with pandemics. From the vaccination limitation by optimal efficiency rates and mismatched effect, antiviral agents are emerged as anti-influenza infection, as the commercial inhibitors of *N*-acetylneuraminidase (NA) are available for oseltamivir and zanamivir on the market. However, the NA inhibitor resistance of H1N1 virus desires a new generation of antiviral drugs.

Influenza virus A and B type viruses comprise two distinct glycoproteins. They are viral spike proteins that are known for hemagglutinin (HA), a sugar receptor-binding protein, and for NA that destroys the binding receptor or cleaves the terminal sialic acid. For HA activation, the inactive precursor HA0 is cleaved to the disulfide-bond HA1 and HA2 by host enzymes. HA1 has an α-SA glycan-binding domain. The two spike proteins of HA and NA are particularly characteristic of influenza virus during the influenza viral life cycle. They directly function in the virus processes, including host interaction, adhesion, host plasma membrane penetration, endosome arrangement, ER-Golgi trafficking, viral particle maturation, viral assembly, and burst-out release, of influenza virus A and B. Among them, influenza type A virus is subtyped into antigenically distinct 16 HA and 9 NA subtypes and thus theoretically 154 subtypes are calculated as possibly emerging viral candidates. The NA spike glycoprotein is enzymatically active sialidase enzyme to cleave NeuAc or SA from viral receptor carbohydrate chains. The NA activity is, therefore, essential for virus budding, cleaving off them, from the host cell membranes. In general, influenza virus NAs are classified into two major groups. Group-1 virus NAs include four different types such as N-1, -4, -5, and -8, while the Group-2 virus NAs include other five types of N-2, -3, -6, -7, and -9 for each NA. Contrary to influenza virus A and B types, the C type influenza virus is known to cause relatively mild respiratory symptoms upon infection in humans and differs from the other influenza virus types of A and B. Interestingly, the host specificity of the influenza virus type C is in its preferred receptor sugar structure, where *N*-acetyl-9-O-NeuAc being the binding substrate. Therefore, the host receptor sugar structure is destroyed by the C type viral enzyme that is the acetylesterase specific for *N*-acetyl-9-O-NeuAc only, but not other substrates such as NA.

Cell surface receptors are directly utilized to bind by a virus and they are generally glycoconjugates such as GSLs, glycoproteins, and proteoglycans. GSLs can directly be used as receptor molecules such as hormones, interleukins, interferons, lymphokines, and cytokines or microbes such as viruses, bacteria, and microbial exotoxins through their carbohydrate glycan structures. Glycan moieties extend into the extracellular outwards, and therefore, virus binding targets the membrane distal carbohydrate and the proximal proteins of membranes. Terminal, internal, or non-terminal carbohydrate regions of GSLs are used as the minimal carbohydrate recognition site. Short GSLs closely located in proximity to the membrane lipids are used as virus receptors. Amino acid sequence around the receptor recognition site defines its binding specificity. Host cell carbohydrate receptor-microbial ligand interactions are low in affinity and are enhanced by multivalency of receptors. Glycan binding is performed by electrostatic forces to negatively charged SA-based carbohydrates or glycans. Such carbohydrate motif is found in glycolipids and glycoproteins.

SA residues are essential for influenza virus receptors [326], but the fine definition of the receptor molecules is not settled down. Sialic acid residues linked to gangliosides are cell entry receptors of many viruses. In pathogenic agents, mumps, influenza, corona, parainfluenza, noro, rota, and DNA tumor viruses use such SA residues linked to gangliosides. Receptor-binding molecules are located on viral envelopes or on the nonenveloped virus surface. In order to de-attach the burst viral particles from host cells, sialidase or sialyl-O-acetyl-esterase is expressed as the receptor-destroying enzymes. The enzymes function to release virus particles from infected cells and protect sialyl-conjugates interfering during viral attachment. The roles of gangliosides in virus entry into target cells remain unclear. SAs are linked in linkage modes of α2,6 or α2,3 glycosidic bonds to a Gal residue and also of α2,8 bonds to a pre-existing internal α2,3-SA. Diverse glycosidic linkages and substitutions are found on the pyranoside ring or SA side linkages produce SA structure diversity. Viruses preferentially bind to SA attached carbohydrates, and this binding specificity determines ranges of virus hosts, tissue-specific tropism of virus, cell type specificity, and viral pathogenic progression. In fact, the influenza A virus of humans and avian prefers to α2,6- and α2,3-SA residues, respectively, and thus the sugar type distribution of each SA linkage type correlates with the glycosidic linkages of the different host cells. The respiratory tract epithelial cells in humans and intestinal epithelial cells of birds express their species-specific SA type for each viral preference [417]. The interactions between viruses and SA linkages of carbohydrate contribute to host range and tissue tropism.

4.10.2 *Attachment, Endocytosis, and Influenza Virus Host Tropism*

From the three influenza virus types, influenza A virus type is a causing agent, frequently raising for tropically threatened pathogenicity in humans and animals. The environmentally primary infection hosts of influenza A virus are avian (birds or chicken) and occasionally transmit to other animal species via changes in SA-recognizing capacity of HA. The changes in host range occur by genetic shift events caused by homologous recombination in swine host. Although the origin of influenza B virus is not clearly known, influenza B virus is considered as a descendant of avian influenza A virus and this type causes mainly respiratory infections in humans, but the detailed mechanism remains unknown yet. During the last two decades, influenza virus A type has been pandemic in specific local areas from Asian countries [418]. The infection of influenza A virus requires its interaction with the cell surface glycan receptors using the surfaced HA and NA. The virus envelope HA and NA are also used for protection in host antibody response, too. Influenza A virus infects the host species with high specificity due to their HA specificity to SA residues. The pandemic influenza A virus globalization spends a hugely social and economic expenditures. Influenza virus A is reported to be particularly susceptible to environmental responses, provoking the genetically homologues mutation through interspecies and eventually spreading out with an avian-swine-human crossed trans-infection. The avian-swine-human crossed trans-infection is accelerated by inter-species hyper-genetic recombination. Therefore, it is assumed that influenza A viruses are specifically diverse in their wide host ranges from avian to humans with widely evolved adaptation in natural selection [419]. Interaction between influenza viral HA and NeuAc residues in carbohydrates causes host infection. Attachment to NeuAc is performed by HA receptor-recognizing proteins as components of viral envelopes or nonenveloping viruses. Changes in the receptor specificities determine virus tropism, specificity, and transmission of viruses. Historically, NeuAc or SA was the first example of the virus receptor (Fig. 4.5) [420]. SAs are modified to the active substrate form, CMP-NeuAc (SA), by a specific CMP-SA synthetase enzyme in nucleus. The sugar nucleotide, CMP-SA translocation to the Golgi apparatus, is mediated by a specific enzyme, CMP-NeuAc translocase, once CMP-SA synthesis is ready, and they are used as the donor substrate of

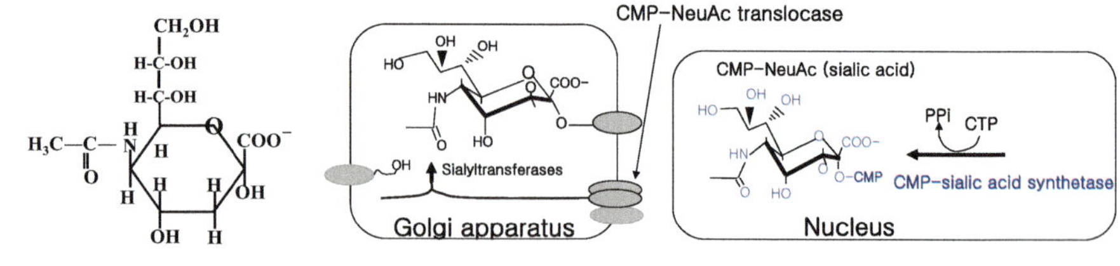

Fig. 4.5 *N*-acetylneuraminic acid (Neu5Ac)/sialic acid (SA) structure and nuclear formation of CMP-SA and translocation into Golgi apparatus

sialyltransferases. Hirst and McClelland and Hare, for the first time, described that influenza virus can hemagglutinate erythrocytes [418, 421]. The NeuAc or SA-recognizing viruses include many animal and human pathogenic viruses including influenza, corona, mumps, noro, parainfluenza, rota, and tumor viruses in orthomyxovirus or paramyxovirus, or etc.

Influenza A virus exhibits each specific species and tissue specificity, which is derived from HA sequence differences, because HA displays the preference of human influenza A for α2,6SA in human and avian virus form for α2,3SA in avian. Through the interchangeable hypermutation susceptibility of the virus, three worldwide influenza pandemics were present with the specificities having H1H1, H2N2, and H3N2 types to date [422]. The first pandemic outbreak of the lethal avian influenza A H5N1 type virus was spread in 1997, and the virus continues to adapt and evolve, yielding global sanitary concerns. During the past decade, a few cases of emerging viral strains were recorded for the 2009 H1N1 swine-origin virus [422] and the H7N9 2013 Chinese virus [423] received much attention from the health controls due to the serious public panic. The avian influenza A virus H5N1 is the lethal and pathogenic influenza virus, and it was considered a powerful threat to domestic animals and humans, although the avian-human inter-mutation is not well understood yet. The recent issue why the pandemic and the worldwide spread of avian influenza H5N1 virus in birds leads to the increased bird-to-human transmission get reached global concern. To prevent the influenza viral respiratory diseases, Flu vaccination is considered an effective strategy to prevent virus infection, but annual reformulation to match antigenic variations is needed for application.

Influenza viruses selectively bind to NeuAc- or SA-linked Gal whereas avian type influenza viruses bind to NeuAc- or SA-linked Gal [424]. The human virus infection is through airway epithelium of the upper respiratory tract. For infection of influenza A viruses via receptor-mediated endocytosis, surface glycoproteins HA, for the first step of virus entry, binds to the terminal NeuAc or SA-linked receptors present in the target cell surfaces through the HA molecule interaction with α-sialoglycoproteins or α-sialoglycolipids [425, 426]. NeuAc or SA bound to gangliosides and glycoproteins are receptors of a variety of viral entries. In complex glycoconjugates, NeuAc or SAα2,3-/SAα2,6-Gal, SAα2,3-/SAα2,6-GalNAc, SAα2,6-GlcNAc, or SAα2,8-SA residue is the sialylated structural patterns. Influenza viruses can not recognize α2-8Neu5Ac in ganglioside GD3 series or polysialyl glycoproteins, but recognize only α2,3- or α2,6-bound NeuAc or SA residues. For example, Neu5Acα2,3/Neu5Acα2,6GalNAc, Neu5Acα2,3/Neu5Acα2,6Gal, and Neu5Acα2,6GlcNAc are the cases [420]. The next step is that NA cleaves the SA-linked receptor present in the target cell surfaces. Some viruses carry sialyl-O-acetyl-esterase instead of NA. The enzyme destroys the receptor to potentiate the infected virus to release from infected host cells.

Mutations in HA lead to the influenza A adaptation ability to the respiratory tract of humans because α2,6-SA is synthesized on columnar epithelial cells of human airway respiratory tract. SA linkage specificity is also observed in papovaviruses with the tropism and pathogenicity. The virus host specificity is derived by the differences in SA- or NeuAc-binding specificity [421, 427], where the human-type receptor has α2,6-linked NeuAc- or Sia-sugars and the avian receptor carries α2,3-linked NeuAc- or Sia-sugars, respectively [428]. Human-type influenza virus-binding NeuAcα2,6Gal carbohydrates are broadly expressed in most of the tissues including duodenum, heart, ileum, lung, liver, and spleen. In contrast, avian influenza virus-binding NeuAcα2,3Gal carbohydrates are found in the distinct tissues such as trachea and lungs. For attachment and entry, human virus has to bind to α2,6SAs, and α2,6-SA cleavage is required for virus release. Therefore, the NeuAc- or SA-sugar receptor-binding specificity is the key player and the most important barrier in H2N2 and transmission [429, 430]. For species transmission of avian virus type to humans, swine plays a key role as the intermediate hosts, because the swine trachea and airway epithelium have both of α2,6-linked NeuAc- or Sia-sugars and α2,3-linked NeuAc- or SA-sugars, respectively. The events of homologous genetic recombination take place between the human-type and avian-type influenza viruses [431, 432].

Respiratory tract epithelial cells of human airway are the first infection line of influenza virus in humans [433]. However, the infectivity of each avian influenza virus to human respiratory track is not well defined yet, and the relationship between the viral infection and transport behavior of NeuAc or SA receptors is not also well explained. In addition, there is no detailed mechanism(s) for the distribution of SA-linked receptors to date. The distribution of avian and human-type SA receptors can indirectly be analyzed in various organs by SNA lectin, which is isolated from *Sambucus nigra*, and MAA-II lectin, which was isolated from *Maackia amurensis* agglutinin (MAA), respectively. Lectins SNA and MAAII recognize the α2,6–/α2,3-NeuAc or SA-containing receptors present in the epithelial cells and organs of the respiratory tract [434, 435]. Basically, human beings synthesize the sugar linkages of NeuAc- or SA-α2,8–/α2,6–/α2,3-Gal residue through its catalytic enzymes such as α2,8, α2,6-, and α2,3-sialyltransferases specific for glycoprotein or glycolipids through the endoplasmic reticulum/cis-media-trans-Golgi apparatus (ER-Golgi system), as shown in Table 4.1. The representatives of NeuAc- or SAα2,3-, α2,6- and α2,8-glucosidic glycoprotein include N-glycan or O-glycans on glycoproteins expressed on the cell surfaces (Fig. 4.6). In glycolipids, the examples of NeuAc- or Sia-α2,3-linked lipids are the ganglioside GM3, which is NeuAc- or Sia-α2,3Galβ1,4Glc-ceramide. Thus, three sugar residues of sialyllactose are exposed as hydrophilic moiety on the cell surfaces of host cells or viral enveloped coats.

Influenza virus attaches its host cells via the viral HA binding to sialylglycans of host cell surfaces. A series of viral life cycle includes endocytosis, endosomal acidification, M2-driven uncoating and HA-driven viral-endosome fusion, and viral genome transportation to nucleus. Influenza virus also undergoes clathrin-mediated endocytosis as well as dynamin-2 GTPase or caveolae-driven endocytosis. Also, the dynamin-independent micropinocytosis also occurs via the receptor tyrosine kinase (RTK) activity. The FGFR2 and FGFR4 with the RTK are involved in influenza virus entry as the first cases, because the influenza virus is associated with the EGFR. The PDGFRβ-GM3-interacted influenza viral endocytosis involves PDGFRβ phosphorylation but the RTK inhibitor Ki8751 inhibits PDGFRβ

Table 4.1 Sialyltransferases for α2,3-, α2,6-, and α2,8-sialyl linkages

Sialyltransferases	Abbreviation
Galβ1,3GalNAcα2,3-ST	ST3Gal I
Galβ1,3GalNAcα2,3-ST(second type)	ST3Gal II
Galβ1,3(4)GlcNAcα2,3-ST	ST3Gal III
Galβ1,4(3)GlcNAcα2,3-ST	ST3Gal IV
GM3 synthase	ST3Gal V
Galβ1,4GlcNAcα2,3-ST	ST3Gal VI
Galβ1,4GlcNAcα2,6-ST	ST6Gal I
Galβ1,4GlcNAcα2,6-ST	ST6Gal II
GalNAcα2,6-ST	ST6GalNAc I
Galβ1,3GalNAcα2,6-ST	ST6GalNAc II
NeuAcα2,3Galβ1,3GalNAcα2,6-ST	ST6GalNAc III
NeuAcα2,3Galβ1,3GalNAcα2,6-ST (second type)	ST6GalNAc IV
GD1α synthase	ST6GalNAc V
GD1α/GT1aα/GQ1bα synthase	ST6GalNAc VI
GD3 synthase	ST8Sia I
Polysialic acid synthase	ST8Sia II
NeuAcα2,3Galβ1,4GlcNAcα2,8-ST	ST8Sia III
Polysialic acid synthase (PST-1)	ST8Sia IV
α2,8-sialyltransferase	ST8Sia V
α2,8-sialyltransferase	ST8Sia VI

phosphorylation. Also, influenza virus neuraminidase cleaves SA residue to yield desialylated PDGFRβ. Virus entry involves the Raf-MEK-Erk signalings. During cell attachment and endocytosis, influenza virus swindles the GM3-associated PDGFRβ signaling cascade. RTK function is modulated by gangliosides in the lipid raft microdomain. Ganglioside species positively or negatively regulate diverse RTKs. Although gangliosides are associated with influenza virus attachment and entry, their intracellular roles after virus endocytosis are not studied. Recently, an RTK inhibitor showed the downstream involvement in the post-entering behavior of the cells [436]. The RTK inhibitor Ki8751 inhibited endocytosis of influenza virus via anti-influenza A and B virus activity in host cells that PDGFRβ-expressed RTK associates with GM3-embedded lipid rafts. Upon influenza virus treatment, viruses are attached to GM3-embedded PDGFRβ-associated endosomal vesicle of the host cells.

For more complex gangliosides, the NeuAc- or SA-containing gangliosides including GM1, GM1b, GM2, GD3, GD1a, GD1b, GT3, GQ3, or related glycolipids are expressed on humans (Fig. 4.6). For examples, the CTxB subunit of *V. cholerae* binds to the GM1, which has the NeuAc- or SAα2,3-linked structures in the enterocytes surface of the intestinal tract [437]. In addition, considering the ABO blood type in humans, the question is raised why are the so-called human type NeuAc or Siaα2,6Gal receptors for influenza virus binding specifically expressed on the human respiratory cells only, but not NeuAc or Siaα2,3Gal receptors? [438]

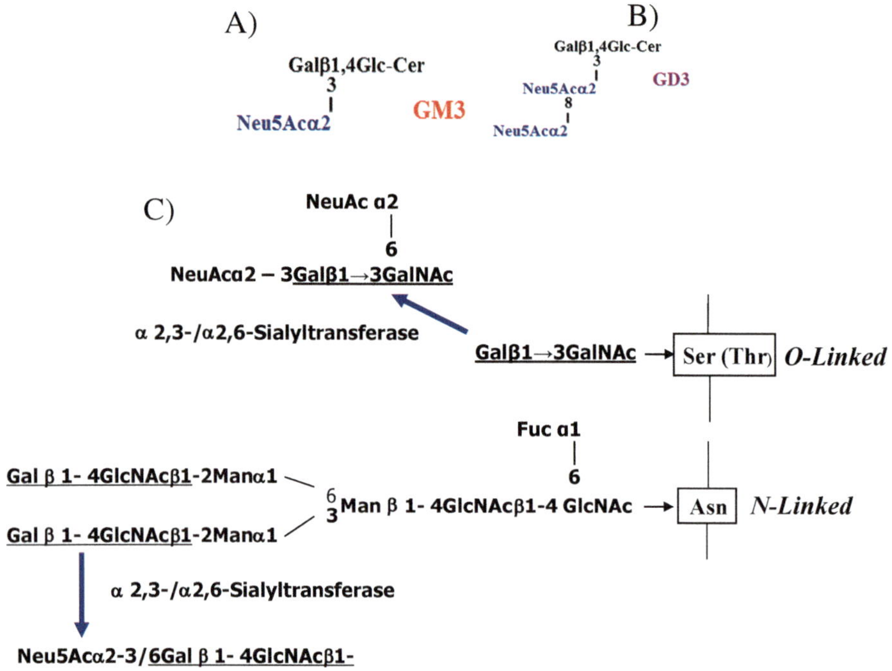

Fig. 4.6 Structures of sialyl α2,3, α2,6, and α2,8 linkages of ganglioside GM3 (**a**) and GD3 (**b**), as well as *N*- and *O*-glycan sugar chains (**c**) of glycoprotein

Furthermore, neutrophil homing and rolling is also associated with normal glycosylation with the expression of carbohydrates with the NeuAc- or SAα2,3-linked structures on glycoproteins or glycolipids [439]. Expression of the sialylated-type 1 or 2 chain with SAα2–3Galβ1,3/4[Fucα1,4/3]GlcNAc structure for sialyl Lewis^{A} (SLA) or SLX is enhanced by α2,3-sialyltransferases of ST3Gal I-VI in those functional cells, because SLA and SLX act for the E-selectin adhesion to endothelium. The avian HAs adaptation to human host cells is thus based on the switching-on affordable for binding specificity from avian virus type of α2,3-SA residues to human virus type α2,6-SA residues present in the upper airway respiratory epithelial cells. The typical SAα2,3-glycan structure represents the NeuNAc-α2,3Galβ1,3/4GlcNAc sequence motif. The SAα2,6 typical structure is the NeuNAc-α 2,6Galβ1,4GlcNAc. The virus binding to sulfatides and neutral glycolipids has been known for influenza A and HIV.

4.10.3 Neuraminidase (NA) Inhibitors as Influenza Virus-Inhibiting Drugs

The NA was isolated as a candidate for influenza virus-targeting drugs. NA-specific inhibitors of zanamivir, oseltamivir, and peramivir are used for the influenza. Oseltamivir was the first drug against influenza. Recently, due to emerging influenza virus strains including H1N1 and avian H5N1 [440], effective drugs are required to cope with influenza viruses. In 2009, peramivir was opened in America against swine-origin H1N1 (A) resistant to oseltamivir. H5N1 virus threatens humanity. Although multiple neuraminidase inhibitors were developed for influenza infection, the rapid occurrence of drug-resistant viral variants needs multifunctional anti-influenza drugs. The resistance issue of influenza virus-inhibiting drugs reflects the development of the new antiviral agents. One anti-entry agent of virus endocytosis is a solution in blocking of virus-host cell recognition in the entry event. At present, the neuraminidase inhibitory drugs have been approved in many countries, but it is mainly administered intravenously, which is very inconvenient for patients. To replace such drugs, recently, many plant-derived polyphenols exert antiviral effects against influenza virus and anti-NAs with strong activities against N2 and N1. For example, the 3,4-dihydroxyphenyl group from caffeic acid was essentially interacting with the NA active site according to the docking analysis, whereas several CA derivatives acted as non-competitive inhibitors. Potent NA inhibitors from caffeic acid derivatives can be designed to cope with influenza virus. CA inhibits and eclipses the multiplication of influenza A virus in vitro, while the progeny virus yield was markedly decreased in the presence of CA [441]. They are diverse in their structures, as shown in flavonoids, xanthones, and diarylheptanoids [442–445]. CA derivatives had potent anti-influenza virus activity and NA inhibitory activity in vitro. The potential pharmacological use of CA or its derivatives is admitted for an antiviral drug against influenza A virus. Unlike the classical structure, these structures are many and act as noncompetitive NA inhibitors, in such a way that they inhibit the NA activity by binding on non-active sites of the enzyme. They could be new candidates for developing new anti-influenza agents. Small natural compounds can be used for new therapeutic agents using fused chemobiological synthesis in NA biology without side effects. Chemobiological fusion science using the natural compounds as the chemical scaffold will create new drugs for influenza. We remember that the discovery of low molecular weight Aspirin with its ingredient acetylsalicylic acid upgraded the human health and welfare during the twenty-first century. A part of the present description has previously been published in the Editorial, *Journal of Glycomics and Lipidomics*, which is not a cited journal.

4.11 GD1a-Porcine Sapelovirus Infection

The word Sapelovirus comes from Simian, Avian, and Porcine entero-specific viruses. The Picornaviridae family consists of 29 genera with nonenveloped and positive-sense ssRNA viruses. Picornaviruses are known to be various infection diseases-causing agents for intestinal, respiratory, neurological, and cardiac diseases in both humans and animals. The genus Sapelovirus is composed of two species of Sapelovirus A that is formerly termed Porcine sapelovirus and Sapelovirus B that is formerly termed simian sapelovirus. Although a third species called Avian sapelovirus known as duck picornavirus TW90A was previously classified, the species is currently reclassified to a family of a new genus termed Anativirus A. The two sapelovirus species are distinguished by several parameters including difference in host species and virus diversity in sequences. In fact, VP1 gene has about 50% homology in amino acid sequences of viral species. The simian type of sapeloviruses, which are suggested to three sero-types, is currently progressing their inter-species recombination, as found in several genomic sequences. The porcine sapelovirus (PSV) is a causative agent of reproduction disorder, diarrhea, pneumonia, and polioencephalomyelitis in pigs. PSV recognizes SAα2,3-linkage on GD1a glycolipids as a receptor, although the role of GD1a is still not clear in viral pathogenesis. However, PSVs are not bound to histo-blood group antigens (HBGAs). PSV belongs to the *Picornaviridae* family, characteristic of small, nonenveloped, single-stranded, and positive-sense RNA genomes. Sapelovirus genus as a picornavirus includes simian, avian, and porcine picornaviruse [446]. PSV is widely distributed with prevalence and the receptor for PSV has recently been defined, as PSV uses SAα2,3-glycan on GD1a for recognition and entry into target cells.

4.12 GD1a/Gal-Cer/HBGA-Murine As Well as Human Norovirus Infection of Caliciviridae Family

Terminally linked SA-Gal or SA-GalNAc residue by an SAα2,3 or SAα2,6 linkage is determined as receptors for viral host range, pathogenesis, and tropism. Enterovirus, reovirus, and rotavirus have their host tissue tropisms of the intestinal tract. Gangliosides having one or more SAs in the α2,3 and α2,8 linkages like GD1a are attachment receptors for Noroviruses (NoVs) during the entry of the virus cycle. NoVs belong to the *Norovirus* genus in the *Caliciviridae* family. As the shape of small, round-structured and ssRNA, positive sense RNA virus with about 7.7 kb length. This infects both animals and humans. Human NoVs are a causative agent of epidemic acute gastroenteritis. No effective vaccine or antiviral therapeutic drugs are developed for human use. They are nonenveloped RNA virus group and thus a NoV capsid protein encapsulates the ssRNA. From lack of this viral culture model, norovirus-like particles have been employed with recombinant capsid protein

(VP1). The single viral protein VP1, comprised of the two shell (S) and protruding (P) regions, is organized to 90 dimers. The P domain has two subdomains of P1 and P2, where the subdomain P2 recognizes its host cells. The HBGAs are known as human NoVs attachment factors or receptors in a strain-dependent way [447, 448], although binding of P-domain to gangliosides is also a basic step for various non-human caliciviruses or non-human noroviruses. Primary infection depends on P domain binding to the HBGA. However, HBGAs always do not account for tropism of all human NoVs and entry into the hosts. HBGAs attached to PM glycolipids or glycoproteins are largely present in surfaces or erythrocytes and mucosal epithelial cells as well as present as free carbohydrate forms of fluids like saliva. Similarly, heparan sulfate and secreted oligosaccharides of human milk are also engaged. NoVs bind to HBGA in all ABO, Lewis, and secreted fluids. Certain human NoVs like Noda485 are not bound to the known HBGAs, implying that HBGAs are not the solely binding and entry targets for human NoVs. Caliciviruses of FCV, MNV, PoSaV, and TV specifically interact with sialylated glycans and HBGAs.

Gangliosides are also another ligand for NoVs in human and murine [327, 447]. Human NoVs recognize GM3 as the highest affinity ligand [410]. Among the GM3 structure, the Sia-Gal-Glc moiety has been demonstrated as the dominant recognition site for the NoV in the ESI-MS assay. The additionally glycosylated gangliosides such as GM1, GM2, GD1b, GD2, or GD3 exhibited reduced recognition capacity, when compared to GM3. The SAα2,3-linkage in ganglioside has been confirmed in norovirus infection. Human Nov P domain directly binds to α2,3-linked SA residue on 3′-sialyllactose (3SL). From saturation nuclear magnetic resonance (MRI), surface plasmon resonance (SPR), and mass spectrometry (MS), ligand-binding epitopes, binding affinity, and stereochemistry in the ganglioside carbohydrate-VP1 interaction have been reported [449]. In addition, GD1a is an attachment and recognition receptor for MNV strains, which was isolated from the brains of immunocompromised mice [35]. Although GM2, asialo-GM1 (named GA1), and GD1a are the normally expressed gangliosides of murine macrophages, MNV-1 strain recognizes a terminal SA-Gal linkage of GD1a and this is the reason why the MNV-1 does not bind to asialo-GM1 or GM1 gangliosides. The lack of ganglio-specific GSLs in murine macrophages through GlcCer synthase inhibitors decreases the binding and infection levels of MNV-1 to murine macrophages. However, the GD1a treatment rescues the defected phenomenon of viral infectivity.

HBGAs attached to membrane proteins or glycolipids on the membrane surfaces of erythrocytes and mucosal epithelial cells are frequently found in saliva or milk. Human NoV-HBGA interactions are observed in ABO or Lewis antigens [450]. Rabbit hemorrhagic disease virus (RHDV) is also a positive-sense ssRNA virus of the Lagovirus genus in the Caliciviridae family [451]. RHDV of caliciviruses targets HBGAs of the host cell for attachment. The lack of HBGA ligands in rabbit exhibits resistance to RHDV infection. RHDV VLPs recognize the A type 2 HBGA glycans, B type 2 HBGA glycans, and H type 2 HBGA carbohydrates expressed on the erythrocytes and epithelial cell surfaces. Some NoVs do not bind any HBGAs, suggesting the HBGA-independent interaction, and claiming that HBGAs are not always the sole receptor for human-type NoVs. Then, GSLs and acidic oligosaccharides have been recognized as human NoV receptors. It was known that galactosylceramide and HBGA glycosphingolipids interact with a VLP of human NoV [447, 452]. VLPs also recognize heparan sulfate, sialyl(S) Le^X, SLe^a, S-lacto-*N*-neotetraose, S-lacto-*N*-tetraose, and S-lacto-*N*-fucopentaose but not 3-′-sialyllactose (SL) and 6SL. Therefore, apart from gangliosides, histo-blood group antigens, α2,3-sialylated chain such as SLe^X, and glycosaminoglycan heparan sulfate are also recognized as the attachment receptors for human NoV strains [452].

4.13 GD3/GM2-Zika Virus Infection

In the genus Flavivirus in the Flaviviridae family, flaviviruses have specific envelop and positive ssRNA. A group of Dengue, Japanese encephalitis, West Nile, yellow fever, and Zika virus are included. Viruses are typically transmitted by arthropod vectors. Zika virus has a specific tropism to neuronal infection causing CNS dysfunction. The neurotropic properties of Flaviviruses are similar to those of Dengue, encephalitis, and West Nile virus like Zika virus. Zika virus (ZIKV) is a causative agent of a global health emergency, implicating the virus as an emergent neuropathological agent in the human nervous system, including fetal and ocular brain diseases, neonatal microcephaly, and adult Guillain-Barré syndrome (GBS). The ZIKV is a positive-sensed and ssRNA *Flavivirus* transmittable to humans via mosquitoes [453].

ZIKV infection leads to the development of autoimmune response, GBS, which occurs in the peripheral nervous system [406]. The GBS symptom basically originates from an infectious agent of bacterium *C. jejuni* and also several others including Epstein–Barr virus (EBV), influenza A virus, hepatitis E virus, *H. influenza*, cytomegalovirus, and *Mycoplasma pneumonia*. Its pathologic mechanism is explained by carbohydrate antigen in peripheral nerve tissue and production of the anti-ganglioside antibodies caused by carbohydrate similarities [454]. ZIKV infection increases in GD3-specific autoantibody, because GD3 is largely expressed in neural SC to maintain the self-renewal.

Are gangliosides related to ZIKV GBS and microcephaly? Currently, the clue is in that the virus obtains host membrane glycolipids and incorporates them. For example, molecular mimicry between pathogenic and neuronal ganglioside antigens is a driving force for GBS upon *Campylobacter* infections [455]. Antibodies to glycolipids (GM1, GA1, GM2, GDla, GDlb, or GQlb) were detected in the onset of GBS. Neurological complications occurred at ZIKV infection are thus caused by autoimmune antibodies, targeting gangliosides. ZIKV easily crosses the placenta barriers because the Zika virus is found in pregnant women's amniotic fluid. In addition, ZIKV infection suppresses human neural progenitor cell proliferation [456]. Autoantibody production against gangliosides such as GD3 in ZIKV infection is interesting. Treg/Th17 imbalance is suggested as the neuropathogenic factor

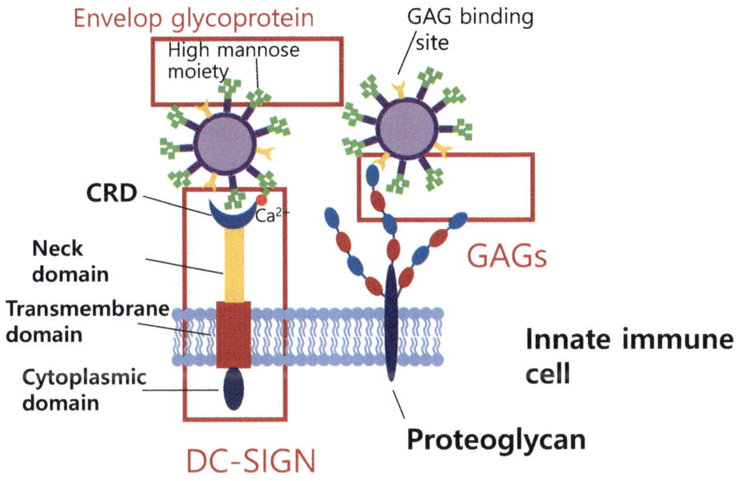

Fig. 4.7 Zika virus-host cell interaction via DC-SIGN and GAGs on host immune cells

during Zika infection [457] with blood–barrier disruption. For GD3 acquisition in Zika virus in the process of viral shedding of infected neural SC that expresses GD3, ZIKV acquires the membrane lipids from the host cells in the course of the virus budding stage. The GD3 is therefore recognized and targeted as an autoantigen by the host immunity because the pathogen-associated molecular pattern (PAMP) can break peripheral tolerance to GD3 [458]. Such produced GD3 autoantibodies suppress the neural cell functions regulated by GD3 functions.

The flaviviruses have similar entry tropism to hosts through GAG-binding site of envelop protein to cell surfaces. Also, high-Man N-glycosylated envelop protein of flaviviruses binds to DC-SIGN of host immune cells. Flavivirus binds to DC-SIGN or GAG on host immune cells (Fig. 4.7). Another marker of ZIKV infection is GM2 ganglioside [459], known for other viruses such as polyomavirus and HIV [460], as GM2 was also elected as an infection biomarker. As a pathogen receptor at membrane's outer layer, the GM2 is important for ZIKV infection with brain tropism.

4.14 GM1/GD1b-Varicella-Zoster Virus (VZV) As Well as GM2-Cytomegalovirus (CMV) Interaction during Infection in GBS

Gangliosides are abundantly expressed in the nervous system, mainly in axons of neuron. Anti-ganglioside antibodies are observed in various autoimmune diseases, including GBS, Miller-Fisher syndrome, chronic idiopathic ataxic neuropathy, multifocal motor neuropathy, and IgM paraproteinemia neuropathy [461–465]. GBS-displaying infection, including *C. jejuni*, varicella-zoster virus (VZV), and Cytomegalovirus (CMV), leads to the production of anti-gangliosides antibodies, allowing many auto-immune diseases. CMV is known as an antecedent infectious agent with its high immortal rate, which phenotypically expresses the demyelination variant of GBS. The fibroblastic cells infected with CMV produce GM2. CMV-infected GBS patients exhibit sensory function deficits with anti-GM2 antibodies as IgM subtype. The known Fisher syndrome is a type of GBS variants having anti-GQ1b antibodies as an IgG type. The produced antibodies reactive for the GSLs cause neuropathies, because such antibodies against GSLs injure and damage motor neurons or sensory neurons, block ion channels, inhibit remyelination, and induce axonal degeneration [465–468].

In GBS patients with chicken pox/VZV infection, IgM types of anti-GM1 antibody as well as anti-GD1b antibody are produced in infected patients [469]. GM1-specific antibody titers and GD1b-specific antibody titers are easily detected in such patients with the GBS after VZV infection, because the antibodies are specific for terminal Galα1,3GalNAc-carrying glycolipids. In Herpes simplex virus and CMV, anti-GM2 antibodies in patients with GBS are frequently detected in the GBS patient sera with CMV [360]. GM2-specific antibodies are also detected in the GBS patients during CMV infection. However, CMV infection does not seem to be related with anti-ganglioside GM2 antibodies. CMV-infected fibroblasts produce GM2 epitopes specific for GM2-specific antibodies. Thus, GM2-specific antibody production in GBS patients infected with GBS is increased by molecular mimics of GM2 and CMV-mediated antigenic epitopes. Antibodies specific for GT1b, GD1b, GD1a, GM3, GM2, and GM1, but not anti-GQ1b antibodies, are detected as IgM and IgG class [470].

4.15 GD1a-Norovirus Interaction

Norovirus (NoV) belongs to the Norovirus genus of the family Caliciviridae. They contain small and round forms of RNA viruses. The virus infects humans and animals [471]. Human NoV (HuNoV) causes epidemic acute gastroenteritis [472]. NoVs are nonenveloped, capsid, ssRNA, positive sensed RNA viruses with 7.7 kb genome size. The capsid VP1 protein of NoVs is a single protein form. NoV-like particles (VLPs) have an icosahedral symmetric structure with VP1. VP1 is classified into P and S domains. The S domain has a specific structure of an interior and icosahedral shell form. But the P domain is featured by the protrusion of dimeric forms [449]. The P domain part is re-classified as P-1 and P-2 subdomains, where the subdomain P-2 occupies the outer surface of capsid protein, which is crucial for the host recognition of virus and immunological recognition of NoVs. It has been implicated that GSLs and SA-based acidic carbohydrates function as human NoV-binding ligands. Binding to gangliosides is crucial for the life cycles of many nonhuman caliciviruses and noroviruses. Human NoVs recognize HBGAs and glycosaminoglycan (GAG) heparan sulfate (HS); α2,3-sialylglycans in the type 2 chain and SLe^x are carbohydrate for the receptor for the attachment and infection of HuNoV [473]. HBGA carbohydrates are attached to membrane glycoproteins or glycolipid. Such HBGA carbohydrates are frequently found on the erythrocytes and mucosal epithelial cells [474]. Human NoVs interact with HBGAs. The NoV-HBGA interaction is found in ABO blood types, Lewis antigens, and secretor/nonsecretor antigen types [475]. The molecular basis of such interactions is studied using NoV P dimers complexed with HBGA carbohydrates. Some NoVs of VA115 type named GI.3 type do not bind to HBGAs, and therefore HBGAs is suggested not to be the sole receptor for infection of human NoVs. GM3 was the highest affinity ligand for the VA387 P and VLP. Interestingly, the affinity for NoV VA387 is relative to HBGA carbohydrate structure [476]. Gangliosides, acidic GSLs, consist of one or multiple SA residues via SAα2,3 and SAα2,8 linkages and are utilized as the receptors of norovirus attachment and infection [452]. HBGA and gangliosides bind to GII.4 human NoV with binding epitopes and affinities. The binding has been demonstrated by 3SL, the GM3 carbohydrate, and B antigen [477]. Recently, human NoVs have been demonstrated to recognize gangliosides using catch-and-release ionization MS (CaR-ESI-MS) tool [471], in the experiment of the human NoV VA387-P (GII.4)-binding GSL.

In murine norovirus (MNV), the GD1a was bound to MNV-1; however, MNV does not bind to GM1 or GA1 named asialo-GM1 [449]. SA residues in gangliosides are a recognition site and receptor for MNV. MNV-1 recognizes SA sequences appeared on cell lines and cultured murine primary macrophages. Recognition of terminal GD1a SA is the initial step in viral infection during the life cycle. Epitope analysis using 3′-sialyllactose suggested the direct recognition of SAα2,3-linkage to the P domain. The human NoV and carbohydrates interaction emphasizes the multivalent binding property of norovirus. Direct recognition of a human NoV GII.4 P dimer to SAα2,3-linkage further supports a functional role of ganglioside-recognition specificity in norovirus. VLPs of GII.4 also bind to HBGA and GalCer [327, 451]. GII NoVs VLP binds to surfaced heparan sulfate of the host cells, while GII.4 VLPs recognize SA-carrying glycans including sialyl Le^X, sialyllacto-N-neotetraose, sialyllacto-N-fucopentaose, and sialyllacto-N-tetraose with similar binding affinities to HBGA glycans [471]. Recognition of GII.4 VLPs to the SA part of sialyl Le^X and sialyl Le^a was demonstrated [478]. Sialyl carbohydrates, but not fucose, including 3′- and 6′-sialyllactose, bind to the VLP, implying that SA-carrying carbohydrates are also used as ligands of human NoVs. SA-containing oligosaccharides are known to interact with certain caliciviruses (CVs) of animals, such as feline calicivirus (FCV), murine NoV 1, and pig sapovirus [327, 479, 480]. However, the ligand status of SA for human sapoviruses as a calicivirus group still remains further evidenced.

Although genus Sapovirus like Norovirus genus, which is belonged to Caliciviruses cause acute and chronic gastroenteritis in animals and humans are also involved in gastroenteritis outbreaks in pediatrics [481]. The genus Sapovirus includes five genogroups G-I/-II/-III/-IV/-V. From them, G-I/G-II/G-IV/G-V groups target human beings for infection and GIII group targets swine species. Although virus-to-host cell recognition via receptors is the initial cycle of infection, the precise receptors for the Sapovirus genus have not been clear yet. However, HBGA is a target molecule present on host cells for Calicivirus attachment. Similarly, Tulane virus, a monkey rhesus calicivirus, also utilizes HBGA antigen as its infection receptor [482]. Virus particles or virions obtained from sapovirus G-I and G-V group of humans do not recognize salivary HBGAs or synthetically related glycans [483]. The noroviruses and sapoviruses are genetically and closely related groups in humans and animals, and the possibility of zoonotic inter-transmission is of concern [484]. The porcine sapovirus recognizes carbohydrates of both SAα2,3-/SAα2,6-linked glycans as entry receptors. Because porcine SaV does not stably agglutinate erythrocytes that carry the SAα2,3 and SAα2,6-linkages, porcine SaV is suggested to stabilize interaction between the sialyl glycans.

On the other hand, FCV as the calicivirus family also recognize GD1a for its receptor. FCV is a group of the calicivirus family and is well studied for its binding and entry. Caliciviruses are classified into the genus lagovirus (LaV), nebovirus, NoV, sapovirus, vesivirus, and recovirus [485]. Only genus NoV and sapovirus group enter the human host. Vesivirus also includes the FCV, a respiratory disease-causing agent in cats. Rabbit hemorrhagic disease virus as a LaV group causes a necrosis-type hepatitis with hemorrhage leakage and mortality. Rhesus macaques Tulane virus and newbury agent-1, a bovine enteric virus named Newbury1, belong to recovirus and nebovirus groups, respectively. The FCV recognizes the airway tract in the upper tract via the α2,6-SA linkage and the junctional adhesion molecule-1, following viral endocytosis [486, 487]. HBGAs, type 2 chain α2,3-sialylglycans (sialyl-Le^x), and heparan sulfate are also known to function as the infection receptor for human NoV [488]. SAs in carbohydrates of host cell glycoproteins and glycolipids are widely used as viral attachment receptor to epithelial cells [481]. Terminally linked SA to the penultimate Gal residue, liked by an SAα2,6 or SAα2,3 linkage, is the general character of SA-containing receptors [475]. SA specifies for virus host range and tissue tropism, as most viruses, bacteria, and toxins are reported to utilize gangliosides.

4.16 nLc4Cer-Dengue Virus Interaction and Non-GSLs-Virus Interaction

Dengue virus (DENV) is a viral disease worldwide with dengue fever (DF) such as capillary leakage and hemorrhagic manifestations [489, 490]. DENV causes morbidity and mortality in the world. It is transmitted by mosquitoes such as *Aedes aegypti* and *A. albopictus* strains. Thus, DENV belongs to a mosquito-carrying Flavivirus, endemic in tropical and subtropical area. Infection with one serotype produces serotype-specific antibodies. Dengue-causing viruses include four antigenic serotype viruses named DENV-1, -2, -3, and -4. As DENV belongs to the Flavivirus genus of Flaviviridae, other Flaviviridae family is also known for Zika, Yellow Fever, Japanese Encephalitis, and West Nile viruses. The virus is an enveloped spherical virion with an 11 kb and positive-sensed and ssRNA strand, consisting of an icosahedral symmetry and a nucleocapsid core [491]. Each serotype is antigenically distinct due to 30% or difference in their amino acid sequences. The RNA genome consists of 10 genes on an open reading frame (ORF), encoding for three capsid, precursor membrane (prM), and envelope protein with the additional seven nonstructural proteins (NS) including NS-1/NS-2a/NS-2b/NS-3/NS-4a/NS-4b/NS. The NS proteins are involved in viral replication and host immune evasion. NS1 is dimeric in early stages of infection and hexameric form of NS1 protein is secreted [492]. The NS1 dimer is present on the lumen side of the ER. NS1 recognition with NS4a and NS4b enhances the viral replication complex formation.

Macrophages and mononuclear phagocytes such as DCs are major targets of dengue virus. The Dengue virus envelope glycoprotein (E) as a class II fusion protein has a receptor binding domain [493]. E protein has three distinct domains. Among them, the domain I is located on the hinge region and domain II having hydrophobic amino acid residues is located in a fusion loop with dimerization capacity. Domain III is host receptor molecule-recognizing region. Dengue virus E protein recognizes lectin and also the surface molecules of the host cells. Surface carbohydrates of host cells recognize dengue virus because the carbohydrate recognition is a key point of DENV propagation. Dengue glycoprotein E binds to the target carbohydrates as receptor molecules including HS GAGs and GSL neolactotetraosylceramide (nLc4Cer) on the host cells [494]. Dengue viruses recognize nLc4Cer of the host cell surface. Dengue virus type 2 binds to neutral glycosphingolipid nLc4Cer, Galβ1,4GlcNAcβ1,3Galβ1,4Glcβ1–1-Cer, on TLC plates [495]. Also, Galβ1,3GalNAcβ1,4Galβ1,4Glcβ1–1-Cer, Gg4Cer, and neutral glycosphingolipid L-3, Gg3Cer, GalNAcβ1,4Galβ1,4Glcβ1–1′Cer have been recognized by the virus to a lesser extent. The virus has no affinity to LacCer. Neutral glycosphingolipids are the binding sites of virus and the β-GlcNAc residue seems to be linked to binding to the host cell surfaces. From the fact that DENV propagates between mosquitoes and humans, GSLs acting as receptors including L-3 and nLc4Cer may support virus transmission [496]. In lectins, DC-SIGN mediates entry of DENV to DCs as the primary target. DC-SIGN binds to N-Glycan at position 67. MBP binds to amino acid positions 67 and/or 153. Mannose receptor is a target site of DENV entry into macrophages [497]. The C-type lectin-like protein CLEC5A (MDL-1) as a macrophage receptor is a receptor for dengue virus [498]. Blockade of CLEC5A can be used in a therapeutic avenue of dengue virus disease (Fig. 4.8).

Host PRR binding triggers type I IFN response. Therefore, the DENV PRRs include the above heparan sulfate, nLc4Cer, DC-SIGN and Man residue, CD14, retinoic acid (RA)-inducible gene I (RIG-I) in the cytosolic region, melanoma differentiation-associated protein 5 (MDA5), and endosomal TLR-3 and TLR-7 [318]. In addition, protein folding effectors like heat shock proteins, HSP70/90, and chaperones also interact with DENV serotype 2 (DENV-2) [499]. Laminin receptor, CD14-associated protein, and uncharacterized proteins also interact with DENV [500], [266]. Immature virus has a protruding prM protein trimer with envelope glycoprotein E. Secondary heterotypic infection causes more severe disease than primary infection by antibody-dependent enhancement (ADE). ADE

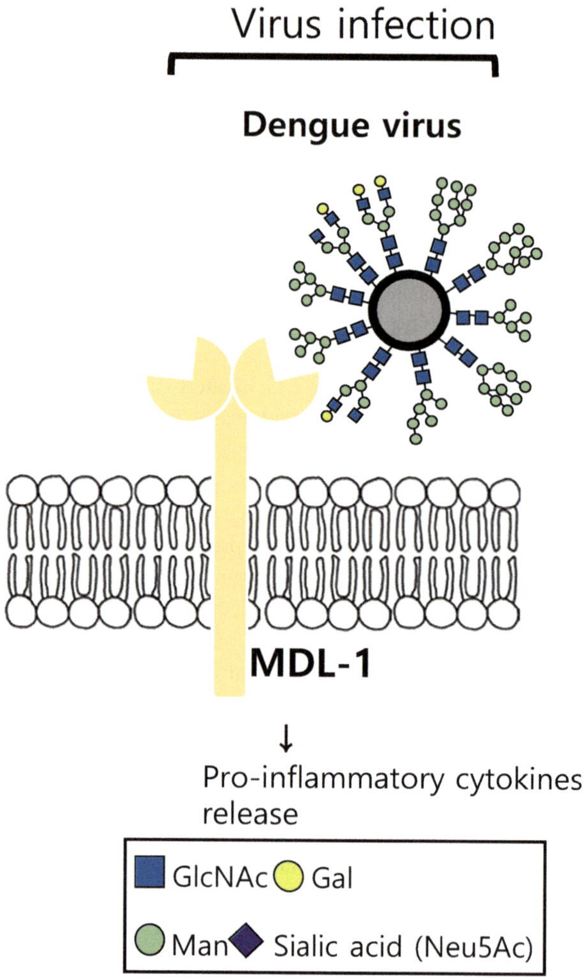

Fig. 4.8 Macrophage receptor, C-type lectin-like protein CLEC5A (MDL-1) as a receptor for dengue virus

displays a virus-antibody complex phagocytosis through FcγIIa receptor [501]. DENV E glycoprotein recognition of the host receptor potentiates virus entry via clathrin-mediated endocytosis. In the Golgi apparatus, the prM is cleaved by furin protease to yield the mature virion and release by exocytosis [502]. RIG-I and MDA5 respond to phosphate-containing RNA and long dsRNA [503]. They translocate to the mitochondrial membrane. Consequently, mitochondrial antiviral signaling (MAVS) protein via the caspase activation recruitment domains (CARD) and MAVS is activated. Thereafter, downstream of TANK-binding kinase 1 (TBK1), IκB kinase-ε (IKKε), phosphorylating IFN regulatory factors (IRF3), and IRF7 are activated to increase the gene expression level of type I IFN. In the TLRs, TLR3 recognizes DENV dsRNA and TLR7 recognizes ssRNA in endosomes. TLR3 phosphorylates TIR-domain-containing adapter-inducing IFNβ, interacting with TNF-receptor-associated factor 3 (TRAF3) and TBK1/IKKε to induce IFNα/β-stimulating genes (ISGs) and chemokines [490]. TLR3 acts synergistically with RIG-I and MDA5 in producing an antiviral state against DENV infection. TLR7 uses the MyD88 signaling. The PRR recognizes cyclic GMP-AMP synthase and stimulator of IFN gene (STING) pathway that recognizes cytoplasmic DNA [504]. DENV damages the mitochondria and mitochondrial DNA activates cGAS-STING. Then, cGAS nucleotidyl transferase releases cyclic GMP-AMP that recognizes STING and induces TBK1, IRF3, and production of type I IFNs. The mtDNA activates endosomal TLR9 that recognizes nonmethylated CpG motifs-containing DNA [505]. In complementation, MBL binds to Man glycans on the DENV surface [506] and induces cleavage of C4 and C2 by MASP-2 and depositing C4b and C2a, forming the C3 convertase, C5 convertase, and C5b-9 membrane attack complex (MAC) to lyse.

Apart from GSLs, heparan sulfate proteoglycans (HSPGs) are used as the primary receptors for dengue virus as well as both HSV type 1 and HSV type 2 but each virus recognizes different structures [498]. Sulfated HSPG is also a receptor for dengue virus [499]. Like DEGV and HSV, adeno-associated virus type 2 (AAV2) binds cell surface HSPGs [500]. Also, as an enveloped RNA virus, Lassa virus (LASV) is an Old World arenavirus known in West African countries. LASV belongs to the family Arenaviridae and is a negative-sense and ssRNA virus. LASV entry needs bindings to O-glycans expressed on the dystroglycan (DG). In addition, the phosphatidylserine (PtdSer)-binding receptors TIM-1, Axl, and Tyro3 as well as CLRs are the DG-independent entry sites for the LASV. TIM-1 potentiates entry. In the absence condition of DG, TIM-1 helps the entry of LASV pseudovirions (Fig. 4.9). Other enveloped viruses such as HCV and Zika virus also use the PtdSer receptors in the viral envelope as GP-independent receptors.

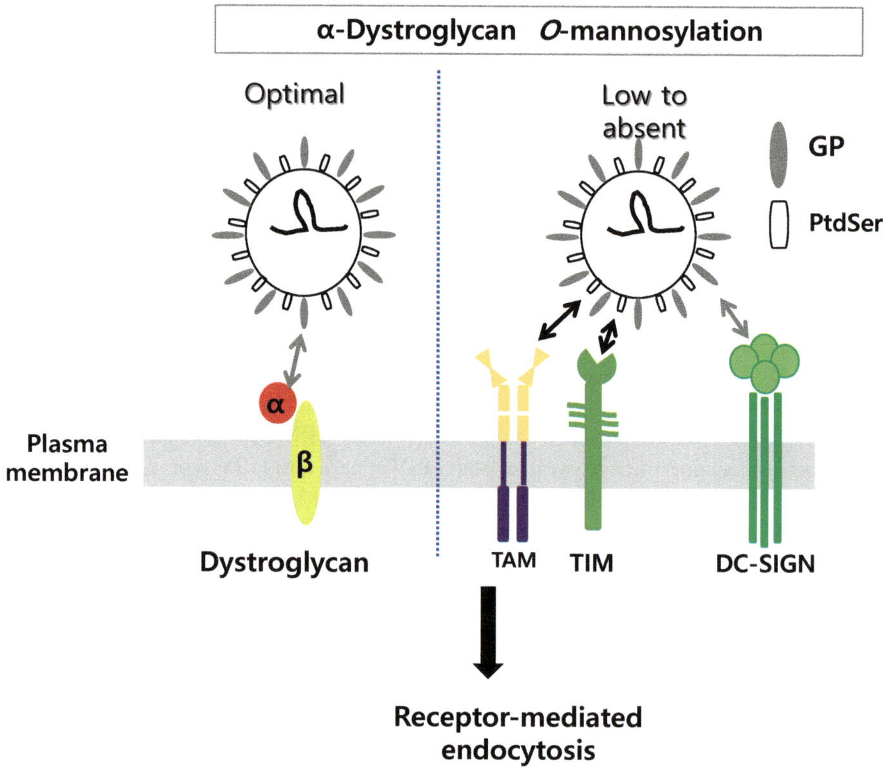

Fig. 4.9 Lassa virus (LASV) binds to O-glycans on the dystroglycan (DG and phosphatidylserine (PtdSer)-binding TIM-1, Axl, and Tyro3 as well as CLRs

Chapter 5
Bacterial Toxin Protein Interaction with Host Cells GSL

Glycosylation events to modify various molecules expressed on cell surfaces are a dynamic process to influence cellular processes of pathogen recognition, leukocyte migration, and tumor immunology. In host cell sides, surface glycosylation varies with species specificity and microbial tropism. In fact, cell surface molecules are normally covered with various glycoconjugates. Glycan structures are modified by GT competition for substrates in the Golgi and alternatively spliced variants. The glycosylation event is modulated at cell type-specific, developmental and differentiation-specific, and environmental stimulus-specific levels. Glycans serve as target molecules for binding of pathogenic microorganisms and virulence toxins. Such recognition indicates the species specificity and pathogen tropism, as studied for viruses. Like viruses, bacteria cause infection as a threat. For example, the bacterial diseases are our major threats due to the rapid infection as well as diagnostic and treating difficulties. The first step in pathogen or toxin infection needs recognition of host cells by the adhesion with carbohydrate receptors on the target cell surfaces through specific and multivalent interactions. The carbohydrate receptors are relatively short chains like oligosaccharides. The cell surface glycans are used for pathogen's entry and innate immunity. Therefore, free glycans such as mucus and breast milk can prevent the attachment of the pathogens or toxins to epithelial cell surfaces [507]. Gangliosides are binding receptors for viral and bacterial toxin. Also, they are cell-cell adhesion molecules and signal transduction modulators. Thus, gangliosides and their GBPs have been explored by many researchers.

GSLs microdomains in the PM are a recognition region or entry receptor for bacterial toxins. For infection, microbes initially bind to specific receptors of host cells or tissue. What is the receptors for each microbe? Most of them are GSLs [508, 509]. GSLs as membrane components with a carbohydrate linked to a ceramide function as receptors for pathogenic agents. Primary receptor functions of GSLs are well known for the bacterial subunit toxins. Gangliosides in the PM with their carbohydrates are exposed to the external region. Bacterial attachment to host cells is a virulence route of pathogens. GSLs occupy a small, typically 5–10% of

C.-H. Kim, *Glycosphingolipids Signaling*,
https://doi.org/10.1007/978-981-15-5807-8_5

total weights, but physiologically and functionally crucial in plasma membrane lipids of eukaryotes. The membrane receptor GSL carbohydrates bind to protein ligands. The binding is affected by GSL lipid moiety and the membrane environment. They represent mechanical stability and biological processes. GSLs form liquid-ordered microdomains for signaling proteins. Bacterial pathogenic protein toxins are virulence factors and bind to carbohydrate receptors expressed on host cell surfaces. Because gangliosides are abundantly expressed in neurons of mammals, they act as anchors or entry sites for exotoxins produced by bacteria. Gangliosides as GSLs are frequently used as cell surface receptors for some bacterial toxins. Representatively, GM1 present in neurons has been redefined as binding receptor, in the early 1970s, for the CTx.

For other cases, tetanus neurotoxin (TeN) and botulinum neurotoxin (BoNT) are also known to recognize the ganglioside b-series of GD1b, GT1b, GD3, and GQ1b. CTx recognizes GM1 with an enough affinity and specificity. CTx, Vero toxin (previously known as Shiga toxin), heat labile toxin (LT)-1, LT-IIa, and LT-IIb of *E. coli* are known to recognize GD1a, GM1, and GD1b. Such bacteria produce extracellular exotoxin and exhibit harmful killing to their host cells by glycan-protein interactions, as GSLs are the entry receptors for the toxins into host cells. Among them, the most well-known example is the GSL-toxin interaction, where the Shiga toxin B subunit (StxB) recognizes the globoside Gb3 as its specific receptor. In terms of the release of pathogenic toxins, a health threatening issue is also raised. From the studies on their binding to target carbohydrates, the anti-adhesion products can be designed for alternative therapeutics.

5.1 Gb3-Shiga Toxin (Vero) Sphingolipid Receptors

The Shiga-like toxins of *S. dysenteriae* and *E. coli* O157:H7 contain a monomeric subunit A and homopentameric subunit B, forming a A1B5 complex. Subunit A has an enzyme activity, but subunit B binds to Gb3 glycolipid present on the cell surface of hosts. All B subunits consist of all three recognition sites, having a total of 15 Gb3-recognition sites. For the subunit B, the first two recognition sites in each subunit directly bind to Gb3. STx or verotoxins (VTs) belong to a group of *E. coli* AB_5 toxin group that induces hemolysis and uremic disease [510]. Verotoxin is also called Shiga-like toxin and verotoxin subunit B binds to Gb3 and potentiates bacterial uptake into the cells. A subunit as an N-glycosidase interacts with the host ribosomes and protein synthesis in host cells is blocked causing the cell death. This disease is a type of renal glomerular nephropathy with anemia and thrombocytopenia. STx produced by enterohemorrhagic *E. coli* strain (EHEC) in the colon region of humans causes fatal death. *S. dysenteriae* infections utilize the Stx 1 of the VT homolog under unsanitary conditions, with Shiga toxin-caused hemolysis and uremic syndrome as an etiological dysentery in children. Colon epithelial cells of Caco-2 and HCT-8 in vitro carry their Stx receptor GSLs such as globotriaosylceramide Gb3 of the Galα1,4Galβ1,4Glcβ1,1-Cer, called CD77 or pk antigen of blood group, and globotetraosylceramide Gb4 of the GalNAcβ1,3Galα1,2Galβ1,4Glcβ1,1-Cer, and lipid rafts. Gb3 and Gb4 ceramide parts are composed of the different fatty acid species of C16:0, C22:0, or C24:0/C24:1 and sphingosine (d18:1) [511]. Their binding specificity is quite broadly extended to the similar structure series of globoseries of Forssman antigen structured with a GalNAcα1,3GalNAcβ1,3Galα1,4Galβ1,4Glcβ-1,1-Cer [512] and globopentaosylceramide Gb5 structured with a Galβ1,3GalNAcβ1,3Galα1,4Galβ1,4Glcβ-1,1-Cer [513].

Shiga toxin type 2 (STx2) is also a member of the AB5 subunits toxins with the A subunit (STxA) which is an N-glycosidase enzyme, acting on the 28S rRNA of 60S ribosomes and inhibiting protein synthesis in host cells. Resultantly, it contributes to proinflammatory signaling known as the ribotoxic stress response [514]. Stx2 binds to Gb3 as a receptor expressed on the endothelial cells, internalizes via a retrograde transport to the Golgi, and induces cellular apoptosis [515]. The five B subunits, which are complex pentameric proteins, bind to Gb3 as a receptor on the target membrane of endothelial cells located in several organs such as kidneys, brain, and intestine [516]. Stx-producing *E. coli* (STEC) produces two Stx proteins of Stx1 and Stx2. Stx2 is known to be largely virulent rather than Stx1 [517].

The binding of subunit with Gb3 is initiated by recognition of the aromatic amino acid residues with the Gal residues of Gb3 [518]. StxB has five identical monomers, and each monomer contains three Gb3-binding sites. The first and third Gb3-binding regions strongly interact with the receptor [519]. As expected, the first and third Gb3-binding regions carry specific amino acid of aromatic Phe30 in the first site, and Trp34 in the third site, keeping the binding to the internal and terminal Gal residues of Gb3, respectively [518]. The amino acid Trp34 residue is located on the α-helix motifs of the StxB monomers. Similar to the GSL-recognizing domain of the HIV gp120 glycoprotein, the GSL-recognizing domain of StxB is thus another example. HIV adhesin gp120 also binds to Gb3. Like VT, lipid moiety regulates the gp120-GSL interaction. Verotoxin 1 (VT1) is homologous to VT2 with the 60% identity. The VT1 and VT2 cause the disease, while VT2 exhibits in vivo severe prognosis. However, VT2 is in vitro less cytotoxic than VT1 in vitro to the host cells [520]. One reasonable point is that VT2 strongly binds Gb3 rather than VT1 in neurological tissue. VT1 and VT2 specifically recognize Gb3 only as a target. However, they also bind to different Gb3 moieties in the PM, suggesting a possibility to recognize aglycone GSL. When a Shiga-like toxin, named Verotoxin B subunit (VTB), has been compared with that of the extracellular region of the B cell CD19 receptor in sequence level, the three different molecules of VT1, VT2, and VT2e B subunits exhibited about 50% similarity with the Gb3-recognition domain in the toxin. In addition, the Aspartic and Glutamic acid residues at the position 16, 17, and 30 in the GSL Gb3-recognizing domain of VTB subunits were conserved in the CD19 sequence at the 30, 32, and 120 positions. Moreover, the aromatic residues Phe30 and Trp34 in the three VTB subunits of VT1, VT2, and VT3 in ShTxB leads to the recognition of hydrophobic stack. They have the commonly conserved sequence in the CD19 at the 122/124 amino acid positions [521]. Thus, the existence of a GSL-recognizing domain in the CD19 indicates a possibility of the Gb3-mediated

CD19 regulation. Thus, it is speculated that the GSL-recognizing domains of ShTxB and gp120 are structurally similar. From the results, the Gal residue located on the GSL core structure is a stacking binding keeper and GSL bioactivity is regulated by the configuration of saccharides [522]. Gal residue, therefore, is abundantly located in GSL carbohydrate glycans and also the most common residue.

Toxin-Gb_3 binding leads to clathrin-dependent internalization and non-dependent internalization. The VT1 in the bilayer binds to Gb_3 and the binding event induces to change the membrane independent of other molecules and VT invites clathrin-dependent retrograde transport through endosomal organelle such as TGN and Golgi-ER. Thereafter, the proteolyzed A1 subunit separated from the whole molecular holotype toxin is delivered to the cytosolic area and inactivates synthetic machinery of polypeptides by ADP ribosylation that de-purinates the 28S RNA molecules associated with the 60S ribosomal subunits [523]. As another action distinguished from the protein synthesis inhibition, toxin binding of Gb3 activates transmembrane src kinase [524].

5.2 Gal-Cer/Gb3/Gb4-Fimbrial Adhesins in Enterotoxigenic *E. coli* (ETEC) and Uropathogenic *E. coli* (UPEC)

Enterotoxigenic *E. coli* (ETEC) causing intestinal disease in humans and animals produces fimbrial adhesins as a host and tissue tropism factor [525]. The type 1 fimbriae are related to the P-fimbriae of *E. coli* in terms of urinary tract infections. P-fimbriae recognize kidney GSLs of the Galα4Gal core structure, whereas type 1 fimbriae recognize the urothelial mannosyl glycoproteins. The F1C fimbriae interact with Gal-Cer and Gb3. Gal-Cer is expressed in the bladder, ureters, and kidney while Gb3 in only in the kidney [526].

On the other hand, the glycan part of GSLs interacts with the fimbriae and F4 ETEC. Among three F4 variants of F4ab, F4ac, and F4ad, the F4ac variant is mostly common. F4 ETEC susceptibility is inherited in a way of the autosomal dominant Mendelian trait mapped on chromosome 13. ETEC strains with F4 fimbriae expression cause neonatal type diarrhea and pig post-weaning diarrhea. Three $F4_{ab}$, $F4_{ac}$, and $F4_{ad}$ F4 fimbriae are found to locate with the major adhesive subunit FaeG. The immunoglobulin-like region of FaeG is the glycan recognition domain. A brush border glycoprotein interacts with the fimbriae F4 [527]. The intestine-located sialoglycoproteins as mucin type O-glycans, named IMTGP-1 and -2, are known as the F4ab and F4ac receptors [528, 529]. A β-linked Gal residue is the recognition site of F4ac fimbriae [530]. Intestinal transferrin (GP74) is also used as a F4ab-specific receptor [531]. The F4ad-binding receptor is an intestinal neutral GSL (IGLad) [532] and terminal β-Gal residue is a binding site in IGLad. Also, neoLactetraosyl-Cer of the Galβ1,4GlcNAcβ1,3Galβ1-4Glcβ1-Cer is the IGLad receptor. The Gal or GlcNAc residue is important for the F4 receptor structure, which is located in the non-reducing end of β-glycosidic linkage [530, 532–534]. Different GSLs such as LacCer, gangliotriaosyl-Cer, gangliotetraosyl-Cer, Gb3, lactotetraosyl-Cer, and lactotetraosyl-Cer [532–535] are also receptors. The fimbriae F4ab and F4ac bind to sulfatide and Gal-Cer, while the fimbriae F4ad does not [536]. However, the fimbriae F4ad recognizes gangliotriaosyl- and gangliotetraosyl-Cers.

In a minor strain of F18-fimbriae-expressing *E. coli* involved in porcine diarrhea and edema, the fimbrial subunit FedF of F18-fimbriae strain adheres to the pig small intestine. Pigs express blood group type A and type O antigenic determinants among ABH group [536] and thus, the blood group B type 1 determinants cannot be bound by the F18-fimbriae-expressing *E. coli* in the pig small intestine. F18-fimbriae-expressing *E. coli* specifically binds to GSLs with ABH group antigens present on type 1 core structure and blood group A antigen in the type 4 structure of heptaglycosyl-Cer. The minimal recognition antigenic epitope is indeed the blood group H structure of type 1 sugar, Fucα1,2Galβ1,3GlcNAc. The optimal recognition epitope site is the terminal Galα1,3- or GalNAcα1,3 present in the blood group B type 1 carbohydrate determinant of sugar structure Galα1,3(Fucα1,2)Galβ1,3GlcNAc and the type 1 blood group A carbohydrate determinant with the sugar structure of GalNAcα1,3(Fucα1,2)-Galβ1,3GlcNAc. From the F18-binding GSLs in the small intestinal epithelial cells with blood group types O and A in pigs, the blood group type O binding is the H type 1 pentaglycosylceramide of Fucα2Galβ3GlcNAc-β3Galβ4Glcβ1Cer. However, the pig having a blood group A type has many F18-recognition GSLs including A type 1 hexaglycosylceramide structure with GalNAcα1,3(Fucα1,2)Galβ1,3GlcNAcβ1,3Galβ1,4Glcβ-1-Cer structure, A type 4 heptaglycosylceramide of GalNAcα1,3(Fucα1,2)Galβ1,3GalNAcβ1,3Galα1,4Galβ1,4Glcβ-1-Cer structure, A type 1 octaglycosylceramide of GalNAcα1,3(Fucα1,2)Galβ1,3GlcNAcβ1,3Galβ1,3GlcNAcβ1,3Galβ1,4Glcβ-1-Cer structure, and repetitively located A type 1 nonaglycosylceramide of GalNAcα1,3(Fucα1,2)Galβ1,3GalNAcα1,3-(Fucα1,2)Galβ1,3GlcNAcβ1,3Galβ1,4Glcβ-1-Cer structure. Therefore, FedF of F18-fimbriae-carrying *E. coli* binds to blood group ABH carbohydrate determinants [537].

Similarly, P-fimbriae of uropathogenic *E. coli* (UPEC) infectable in urinary tract in humans are the adhesion molecule to the distal or internal Galα1-4Gal moieties on uroepithelium. PapG adhesin is the Galα1,4Gal-recognition molecule with the three PapG-I, -II, and -III adhesins to host cell GSLs, as confirmed in the *E. coli* UTI and urine *E. coli* strains. Gb3Cer of Galα1,4Galβ1,4Glcβ-1,1-Cer structure, Gb4Cer of GalNAcβ1,3Galα1,4Galβ1,4Glcβ-1,1-Cer structure, and Forssman antigenic GalNAcα1,3GalNAcβ1,3Galα1,4Galβ1,4Glcβ-1,1-Cer structure are bound to the PapG-I adhesin molecule. However, preferences of the two PapG-II and -III adhesins were increased depending on glycan lengths shown in Gb3Cer, Gb4Cer, and Forssman antigen. As they agglutinate erythrocytes, the PapG-I adhesin and PapG-II adhesin prefer Gb3Cer and Gb4Cer, respectively. PapG-III prefers Forssman [538]. The P-fimbriae-globoseries GSL adhesion leads to the anti-adhesion agents applicable for therapeutics against the related infectious diseases.

P-fimbriae in the UPEC strains are expressed by the operon-regulated pap (pyelonephritis-associated pili) genes via the chaperone-usher assembly [539]. The P-fimbria contains two distinct regions of a distal tip and proximal rod. The distal tip is the glycan-recognition motif for Pap-G adhesin [539]. One Pap-G, 1 Pap-F, 5–10 Pap-E, and 1 Pap-K molecules are assembled to form a complex. Pap-C spans the out membrane, while Pap-H is terminally located on the periplasmic space. Galα1-4Gal-carrier GSLs were initially the binding site of the tip PapG adhesin of UPEC [540]. The three PapG-I, -II and -III adhesins are the main adhesins produced by UPEC. Their evolutionary differences between tissue-specific adhesions were reported [541]. The P-fimbriae-globoseries GSL adhesion leads to the future anti-adhesion agents applicable for therapeutics against the infection diseases.

5.3 GD1a/GD1b/GT1b-*Salmonella enteritidis* Flagellin (FliC)

Salmonella enteritidis flagellin (FliC) induces defensin production as an antimicrobial peptide for host protection at mucosal surfaces. Gangliosides are also used as co-receptors of FliC with TLR-5 on the host cells. Exogenous ganglioside GD1a increased the FliC efficiency on defensin expression [542]. Antimicrobial peptides like defensin are host defense molecules against bacteria, fungi, and viruses. Gram-negative *Salmonella* species cause human enteric fever and gastroenteritis. *S. enteritidis* or *S. typhimurium* flagellin (FliC) increased defensin-like protecting proteins via NF-κB activation [543]. Gram-negative and Gram-positive bacteria-produced flagellin utilizes TLR-5 as a binding receptor. Therefore, TLR-5 recognizes *S. enteritidis* flagellin FliC. Asialo-GM1, GM1, and GD1a also recognize *P. aeruginosa*-produced flagellin [544]. Therefore, GT1b, GD1b, and GD1a are co-receptors for FliC during TLR5 interaction. Among them, asialo-GM1 is a *P. aeruginosa* flagellin receptor in epithelial cells [545]. TLR5 activates *S. typhimurium* flagellin-mediated p38 phosphorylation rather than through asialo-GM1.

5.4 Neu5Ac-*Salmonella enterica* Serovars Typhi (S. Typhi)

Human pathogenic Salmonella Typhi cause typhoid fever. Another type of *S. enterica* serovar Paratyphi A (*S. paratyphi*) is a recent agent. Typhoid fever is systemically progressed during disease manifestation, while other *S. enterica* serovars such as *S. typhimurium* is causative for illness with limited gastroenteritis [546]. *Salmonella typhi* typhoid toxin is a key virulence factor of the bacterial human adaptation. *S. typhi* and *S. paratyphi* specifically produce typical typhoid toxins, the AB-family toxin, whereas non-typhoidal Salmonella enterica serovars do not produce such toxin [547]. Unlike other known AB toxin family members, typhoid toxin is an A2B5 structure with two enzyme A subunits of heterogeneous PltA and CdtB and a homopentameric form of identical B subunit made by PltB [548]. PltB B subunit binds to the epithelial cell surface glycoproteins podocalyxin 1 or the myelocytic cell surface CD45. Typhoid toxin specifically binds to human glycans [549] with specificity to Neu5Ac, but not to NeuGc, indicating the human-binding specificity. NeuGc-expressing cells or animals are not sensitive to typhoid toxin. Another ArtAB is expressed in the broad host-specific *S. typhimurium* and the ArtB binding of receptor is different from that of the human-adapted typhoid toxin. ArtB can bind to sialylated glycans structured with Neu5Acα2–3Galβ1–3/–4Glc/GlcNAc and Neu5Acα2–6Gal/GalNAc, which typhoid toxin did not [550]. In contrast, ArtB can also bind to Neu5Gc-carrying glycans. S. Typhi's typhoid toxin secretion appears in the S. Typhi-containing vacuole (SCV) by a specific transport system [551] and packaged vesicles transport to the extracellular space [547]. Typhoid toxin is sorted to vesicle carriers by B subunit and PltB interacts with sialylglycan packaging receptors on the SCV for the toxin-carrier complex formation [552]. Cells negative for N-glycosylation, specific ganglioside synthesis, or terminal Neu5Ac termination cannot transport such typhoid toxin. Therefore, the intracellular bacterial exotoxin utilizes host pathways for packaging and export.

Gangliosides also lead to typhoid toxin sorting process because gangliosides-deficient cells are not packable and exportable for typhoid toxin; gangliosides are involved in the endocytosis sorting of toxins and viruses via the retrograde transport system. Although the N-glycosylation elimination disrupted typhoid toxin export, typhoid toxin is alternatively sorted by NeuAc residues of gangliosides [549], as gangliosides function in the vesicle transportation of viruses and certain bacterial toxins [553, 554]. GM3 synthase (ST3Gal5)-transfected host cells exhibit GM3 production, the precursor of all glycosylated gangliosides and the increased Typhoid toxin export. Silencing of B3GALT4, a β1,3-Galactosyltransferase-4 for the complex gangliosides, increases typhoid toxin export. Thus, the less complex gangliosides are associated with more efficient typhoid toxin sorting. Gangliosides contribute to typhoid toxin exports [550].

5.5 GM1-*E. coli* Heat-Labile Toxin (LT) and *Vibrio cholerae* CTx

Plasma membrane organization is a functional requirement for pathogenic microbes and toxins. For example, *V. cholerae* toxin STs or pathogenic *E. coli* toxins are included [555–557]. They cause diarrhea. *V. cholerae* CTx is the causing factor for the severe diarrhea symptom named "disease cholera," acting in the small intestinal epithelium of hosts. There is no relevant therapy yet but with only electrolyte management to reduce mortality. The CTx first binds to intestinal epithelial host cells. LT type I and CTx are the currently known and the most virulent factors for massive secretory diarrhea in humans, and they are secreted by *V. cholerae* and toxigenic *E. coli*, inducing massive and severe secretory diarrhea. Many toxins belong to the AB5 group with a toxic enzymatic A subunit (CTA) and receptor-recognizing five B subunits (CTB). CTx is so-called AB5 toxin type having a single CTA and five homopentameric CTBs. It is functionally similar to a class of enterotoxins including *E. coli* LT as well as pertussis toxin, diphtheria toxin, shigella toxin, and *P. aeruginosa* exotoxin A. CTx, STx, shiga-like toxin, *E. coli* LT, and pertussis toxin are the representative toxins. CT and LT are similar in AB5 structures, but different in receptor binding specificity [558]. The GM1 carbohydrate of Galβ1,3GalNAcβ1(NeuAcα2,3)4Galβ1,4Glc-ceramide structure is the only receptor for CTx, but LacNAc located on the poly-N-GlcNAc sequences of glycoproteins [559, 560] or paragloboside of Galβ1,4GlcNAcβ1,Galβ1,4Glc-Cer structure [561] is an alternative LT receptor. Because the local microenvironment of GSLs affects carbohydrate presentation for ligand binding, GSL-protein binding is controlled by GSL fatty acid and ceramide contents. Different protein ligands differentially bind to the same receptor GSL, depending on cholesterol, because cholesterol maintains membrane structures with sphingolipids.

CTx resembles heat-labile enterotoxin in its function, structure, and immune response, where they equally use GM1 as the surface receptor. Each B subunit in LTB (Mr ~ 12,000) is compared to the CTB subunit (Mr 11,600). LTs are closely related to CTx. In addition, both LTs and CTx's nontoxic B subunits similarly elicit mucosal and systemic responses. The *E. coli* LT family is divided into two major groups of type I LT (mainly termed LT-I) and type II LT (named LT-II). LT-I and LT-II enterotoxins share similar structure properties. To date, several LT-I variants are known for LTh-I and LTp-I, whereas the known three LT-II variants such as LT-IIa type, LT-IIb type, and LT-IIc type are antigenically independent. Both LT-I type and LT-II type consist of a typical A subunit bearing ADP-ribosylase enzyme and five B subunits. Subunit A enzyme is composed of A1 and A2 subunit. Upon B pentamer interaction with A2 subunit, the B pentamer is ready to form a pore. The B subunits of the LTs have ganglioside-binding domains exposed on the surface side of host cells of mammals. For example, LT-IB recognizes GM1, while LT-IIaB recognizes GD1b, GD1a, and GM1 as well as TLR-2. The LTs are characteristically immunogenic upon interaction with APCs and T cells, eliciting various phenotype changes including cytokine profiles, co-stimulatory molecule expression, antigen presentation, and T cell expansion.

CTx and LT type I are comprised of 5 identical B subunits with a molecular weight of 11 kDa and one A subunit, which has two A1/A2 domains. The pentamer form of B subunit with a molecular weight of 55 kDa binds to GM1 at the apical region of the host cell surface. For secretory diarrhea, the toxin initially recognizes GM1 at the apical epithelial surfaces polarized and enter the target cells via receptor-mediated endocytosis. Finally, they migrate via a retrograde manner to the Golgi cisternae trafficking to the inner organelle, ER, while the A1 subunit peptide of molecular weight 22 kDa is enzymically activated. GM1 (MW, 1564 Da) binds to the CTB and heat-labile enterotoxin. CTB and GM1 are utilized to assess the minimum concentration of toxin. One dimeric CTA subunit (Mr ~ 27,400) and five identical CTB subunits that each CTB has a molecular weight MW11,600 and bear a binding site for its GM1 membrane receptor. The pentameric CTB binds to a GM1 receptor in the PM [562]. The receptor-toxin complex is entered via endocytosis and delivered to the ER [563] via retrograde transport. The A subunit is S-S linked in the loop region between amino acid Cys-187 and Cys-199. The enzymatic A1 peptide, CT-A, is produced by *V. cholerae* protease or gut protease. The S-S bond in CTA localized in the ER lumen is modified to the reduced form of -SH groups and CTx-A1 dissociates from the toxin [564]. CT-A1 thereafter retro-translocates to the cytoplasmic areas by the ER-associated degradation pathway [565]. CTx is, therefore, an ADP ribosyl transferase enzyme [566]. The enzyme catalyzes ribosylation reaction event in the activating the GTP-binding protein's α subunit associated in a heterotrimeric form, which the ribosylated activating α-subunit stimulates the associated enzyme, adenylate cyclase, to activate the CFTR known as a chloride transporter protein, to lose the intracellular water molecules, displaying for the distinct cholerae-type disease. Mechanistically, CT-A1 forms an active complex with an ADP ribosylation factor (ARF). The complex ribosylates the ADP to the G protein α-subunit, increasing cyclic AMP and eliciting Cl^- secretion [567]. CTB subunit can be used for mucosal adjuvant immunity via GM1-involved signal transduction for clinical trials. The similar adjuvant property has been expanded to ganglioside binding *E. coli* heat LT because GM1 binding is crucial for their immunomodulation. How GM1 invites lipid transportation is not elucidated. Pentameric B subunit recognition of GM1 may stimulate host cell membrane potentials, as observed in SV40 and Shiga toxin.

LTs act directly as an adjuvant in the innate immune responses to increase the antibody response to Ag through B cells activation by cytokines produced by innate immune DCs and macrophages. LTs effects exerted on the APC and T cells are elicited through each B subunit interaction with each receptor [568]. GD1b and GM1 are microdomain lipid raft constituents and TLR-2 is shown to traffic to lipid rafts when it binds target ligands. LT-IB or CTB binding in B and T cells [569] induces capping of cell PM receptors and allows redistribution of signaling receptors and cytoskeleton. LT-IB binds to GM1 and activates MHC-II-expressing cells. The LT-IB and LT-IIaB recognize the surfaced GM1 and GD1b of target cells as their receptors, respectively. LT-IB-GM1 binding on B cells activates PI3K and MAP/ERK signaling and MHC-II and CD25 expressions with other costimulatory molecules [570]. LT-IIaB-GD1b-TLR-2 binding on monocytes induces NF-kB-mediated pro-inflammatory cytokine productions [568]. B subunits contribute to the formation of cellular clustering with B and T lymphocytes even in LT-IIaB presence. This B cell clustering is formed by ganglioside. LT-IB elicits expression of IL-2/IFN-γ/CD25/B7–2/CD25 and MHC-II in B cells. In fact, host immune B cells treated with LT-IB elicit the enhanced expression of a variety of downstreamed molecules such as B7–2 (CD86), CD40, ICAM-1 (CD54), and IL-2Rα (CD25) [570]. The LT-IB elicits the expression of IL-2Rα (CD25) in B-/CD4+ T cells and mucosal and antibody productions. LT-IB also elicits cytokine secretion in DCs. LT-IB-treated splenic cells increase IL-5 and IL-4 expression levels but decrease IFN-γ level. The LT-IB induction of anti-inflammatory Th2 cytokine influences immune diseases. Thus, adjuvant function of LT-IB is attractive to its immunogenicity. The mechanistic explanation of LT-IB roles is based on its affinity binding to GM1 present on APC surface and effective presentation to MHC-II and MHC-I. In contrast to LT-IB, LT-IIaB recognizes TLR-2 and GD1b with high affinity and increases the inflammatory IL-1β/TNF-α/IL-6/IL-8 levels in innate immune cells [571]. The interaction between LT-IIab and TLR-2/GD1b on DCs elicits DCs migration in the nasal mucosal region via CCR7 upregulation, Ag presentation, and maturation coupled upregulation of CD40, CD80, and CD86. In addition, LT-IIaB elicits Ag-bound CD4+ T cells proliferation and consequent IgG and IgA production upon certain Ag immunization [568]. Therefore, LT-IIaB elicits proinflammatory immune responses.

The representatively studied B subunits of the LT-IB subunit (termed LT-IB) and LT-IIa B subunit (termed LT-IIaB) act as immune adjuvant components. In the LTs B subunits role in B cells, B cells are clustered by gangliosides not by TLR-2. GM1 ganglioside forms cell clustering with LT-IB and GD1b ganglioside with LT-IIaB as well as GD1a ganglioside for LT-IIbB mediates clustering of cells. Immunogenic and adjuvant LTs come from the specific binding to their receptors present on APCs and it reflects their possibilities as adjuvants with soluble antigens. Clustering of the APCs with T cells implies for the immunogenicity of Ags. Gangliosides regulate signaling of cytoskeleton molecules, as explained in CTB-GM1 microdomain lipid rafts formed in Jurkat T cells toward adhesion, integrin-ligand binding, and cytoskeleton redistribution [569]. Then, consequently, the T cells are resistant to detach and increase F actin elongation and suppress T cell chemotaxis and migration. The leukocyte function-associated antigen-1 (LFA-1) and LFA-1 ligand CD54 present in the B cell and T cell surfaces mediate their clustering by CTx binding to T cell lipid rafts. Like this, the LT-IB binding to B cells induces expression of co-stimulatory CD54 protein [571]. T cells and APCs are clustered by LT-IB or LT-IIaB-ganglioside interaction. The B subunits induce T cell division and IL-2 synthesis of T and B cells. T cells are ready to grow when they interact with the APCs. Cells fall in stop step at the G1 and S stage of cell cycle, causing growth inhibition of the cells but cell size enlargement. The T cell mitosis blockage involves IL-2 release during T cell hybrid formation with B cells. For the mechanism underlying the LT B subunits influence T cell mitosis and IL-2 release upon antigenic stimulation; LTB crosslinks membrane gangliosides or TLR-2 and consequently rearrange signaling proteins. LT-IB/LT-IIaB enhanced antigen presentation for foreign antigens such as ovalbumin after BCR uptake resembles the T cell hybrids action to respond against antigens. For example, LT-IB and HIV gag p24 protein injection in mouse blocks the gag p24-responding T cell division [572]. The BCR-Ag linking undergoes its translocation into GM1-containing lipid rafts, eliciting signaling to induce B cells through the BCR-Ag complex. BCR translocation stimulates B cells-T cells adhesion to cause the T cell mitosis blockage and IL-2 release. LT-IIaB-GD1b and TLR-2 interactions indicate the adjuvant function. LT-IIbB enhances B-pentamer-ganglioside-TLR-2 complex formation on B cells. BCR linking with MHC-II and CD40 immobilizes Ca^{2+}. However, the BCR-CD40 linking declines MHC-II-driven intracellular Ca^{2+} mobilization. T cell hybrids also produce IL-2 upon mitogen stimulation. B cell stimulation by LT-IB indicates T cell stimulation through IL-2 production and a blockage of cell division of T cells. Interestingly, LT-IIaB recognizes GD1b and TLR-2 [568]. B subunit forms in the LT-IB and LT-IIaB trigger clustering, T cell growth, IL-2 release, and MHC-II behavior in innate immune cells like macrophages as an essential process. Therefore, microbial agents such as LTB may delay the immune response in body.

5.6 GD1a-Coreceptor for TLR-2 Signaling and *Escherichia coli* Enterotoxin Type IIb (LT-IIb-B5)

TLRs are known as the mammalian homologues of Drosophila Toll receptor and function as pattern recognition receptor sensors of pathogenic agents in order to induce inflammatory and immune responses [573]. TLRs contain the cytoplasmic Toll/IL-1R domain. Totally, 13 TLRs are known in mice and 10 in humans with several defined ligands [574, 575]. TLR activation is involved in signaling of downstream pathways of inflammation-associated genes [576]. TLR expression is cell-type specific and modulated by cell state and strictly regulated. GD1a is a coreceptor of TLR2. GD1a and LT-IIb-B5 complex strongly binds to TLR2 compared to the LT-IIb-B5 or GD1a [577]. The LT-IIb-B5 subunit B stimulates TLR2 signaling [578]. Type IIb, type I, and type IIa all are an AB5 form with a toxic A subunit and a ganglioside-recognition B5 subunit. LT-IIb-B5 recognizes GD1a [579, 580] as a TLR2 coreceptor. TLR2 interacts with GD1a-recognition subunit of LT-IIb-B5 and the LT-IIb-B5 increases in TLR2 recognition activity with GD1a. LT-IIb-B5 stimulates recruitments of TLR-2 and TLR-1 to lipid rafts to cluster with GD1a, facilitating TLR2 signaling. GD1a cooperates with TLR2 as a heterogeneously complexed receptor for LT-IIb-B5 activation of the host cells. GD1a brings LT-IIb-B5 with TLR2 in PM. LT-IIb-B5 cell activation needs GD1a and TLR2 co-expression. LT-IIb-B5 activation requires Toll-IL-1R domain-containing adaptor protein known as TIRAP colocalized with TLR2 and GD1a in the cells. GD1a displays TLR2 coreceptor role by recruitment of the ligand, binding, and activation of TLR2. Thus, GDa1 is TLR2 coreceptor. The co-receptor concept is a recent trend in TLR4 or TLR2 with accessory adaptors or coreceptors to lead to signaling. TLR4 alone does not induce a strong innate immunity against LPS but needs MD-2 and CD14. In addition to them, unknown molecules in the LPS-binding complex have been suggested to involve in TLR4-mediated signaling [581]. In another example, TLR2 specific for microbial lipoproteins collaborates with TLR-1 or TLR-6 as collaborators and with coreceptors of CD14 or CD36. Then, the forms of TLR-2/-1 or TLR-2/-6 complex can activate the host cells [582]. The TLR complexes with coreceptors can discriminate microbial antigens to perform the host defense and response as repertoire.

Apart from GD1a, asialo-GM1 collaborates with TLR-2 or TLR-5 to increase IL-8 expression level in epithelial cells of humans [583]. Gangliosides are known to bind diverse microbial cell wall structures by PRRs. With the *E. coli* LT-IIb toxin, ganglioside GD1a also binds to the flagellin produced by *P. aeruginosa* strain [544] or Polyomavirus [584]. In a similar fashion, GM1 and GD1b bind to the SV40 [584] and the TeNT fragment C [585]. The *P. aeruginosa* flagellin or some Gram-negative bacterial flagellins induce GSLs-mediated downstream signaling of TLRs [542, 586]. The coreceptor concept is not restricted to TLR only but expanded to the HIV-1 gp120 that recruits CXCR4 associated with GM3 to form a complex receptor to infect T cells [587]. In addition, GM1 is a coreceptor with FGFR upon FGF2 induction [224].

5.7 GD3/GD1a/GD1b/GT1b/GQ1b (b-Series Ganglioside)-*Clostridium botulinum* Neurotoxins (BoNT) and *Clostridium tetani* Tetanus Neurotoxin (TeNT)

The poisonous botulinum neurotoxins recognize their target cells in hosts by a specific binding affinity through binding to two surface receptors expressed on neuronal cells. Ganglioside carbohydrate motifs are the first step co-receptors for the protein complex formation with a membrane protein receptor. The most well-studied neurotoxins currently are botulinum neurotoxins (BoNTs) and tetanus neurotoxin (TeNT), which cause the tetanus diseases and paralytic botulism diseases, respectively. Species of *Clostridium* including *C. botulinum* and *C. tetani* produce neurotoxins of BoNT [588] and tetanus TeNT [589], which mainly bind to the ganglioside b-series GQ1b, GT1b, GD3, and GD1b among gangliosides. For botulinum neurotoxins BoNTs and TeNTs-caused paralysis of botulism and tetanus, the molecular binding of toxins to gangliosides is necessary as the initial step of entry into neuronal cells and binding to a co-receptor protein is followed. Each neurotoxin recognizes the SA-containing carbohydrate motif of gangliosides.

The BoNT serotypes A–F bind to specific GT1b, GQ1b, GD1a, and GD1b and the minimal binding site has been determined. The minimal GalNAcβ1,4Gal carbohydrate sequence present in the asialo-GM2 and asialo-GM1 has been suggested to be crucial for adhesion for many pathogens. Such oligosialyl gangliosides recruit and concentrate TeNT and BoNT on the cellular PM of host cells. Toxin-ganglioside interaction and then consequent binding to protein co-receptors lead to the toxin endocytosis. The ganglioside recognition to various BoNT types including BoNT-A type, -B type, –E type, -F type, and -G type as well as TeNT type is displayed in a conserved sequence responsible for ganglioside-recognition pocket located in the 25 kDa domain in the C-terminal region. However, other BoNT-C type, BoNT-DC type, and BoNT-D type contain two distinct ganglioside recognition motifs [590]. Neurotoxins recognize the carbohydrate motif of gangliosides. *C. botulinum* expresses a binary toxin C2 with an active enzymatic C2I and a binding C2II [591]. C2I is a mono-ADP-ribosylase, working for G-actin at amino acid Arg-177 residue, converting G-actin to a capping protein. This process consequently blocks polymerization of actin filaments. The targeted actin filament is depolymerized with cell rounding [511]. Thus, C2 toxin is classified to a binary actin-ADP-ribosylation toxin group. The actin-ADP-ribosylating toxin family is also known for *C. perfringens* iota-toxin (ADP-ribosyltransferase), *C. spiroforme* iota-like toxin (ADP-ribosyltransferase), and *C. difficile* ADP-ribosyltransferase as well as *B. cereus* insecticidal protein [592].

C. botulinum BoNTs block release acetylcholine through affinity binding to neurons. Seven *C. botulinum* A–G strains exhibit their different antigenic specificities for each produced neurotoxin. BoNT-B type, D type, F type, and G types specifically degrade synaptic membrane proteins in vesicles. The other BoNT-A type, -C type, and -E type degrade specifically the presynaptic membrane proteins, indicating different specificities. In addition, BoNT-A type and -E type degrade the 25-kDa synaptosome protein, termed SNAP-25, whereas BoNT-C cleaves syntaxin. The H-chain has two domains to transport the L-chain to the neuronal cytosols. BoNT binding to the neuromuscular junction elicits neurotoxicity. BoNT known as a botulism maker is one of the most toxic substances with seven distinct serotypes A to G. Among them, BoNT-A and -B are therapeutic for human diseases including cervical dystonia and myofascial pain. Recently, they have been used in cosmetic treatments [593]. BoNT toxins consist of three domains of light chain (LC), translocation region (HN) and binding region (HC). The H-chain amino-terminal domain transfers the L-chain located in the lumen to the cytosols through endocytosis. The H-chain carboxy-terminal domain recognizes the cholinergic neurons. The LC is an enzyme having Zn-protease activity for the toxin-targeting N-ethylmaleimide-sensitive factor (NSF) attachment protein receptor (SNARE) present as a soluble form. The LC as a Zn^{2+}-requiring endopeptidase specifically cleaves SNARE, triggering disruption of the synaptic vesicle fusion. Because SNARE protein functions in synaptic vesicle exocytosis, the BoNT-mediated cleavage of SNARE molecule blocks release of acetylcholines at the motor neurons in neuromuscular synapse. BoNT toxins are primarily generated in a monomeric protein form with a MW 150 kDa and converted to a light LC with a 50 kDa and a heavy HC with a 100 kDa polypeptides. The two LC and HC are linked via the disulfide bond. A 50 kDa HN is the LC transport protein and a 50 kDa HC is a nerve receptor-binding protein, respectively. BoNTs specifically bind the two known neuronal receptors of gangliosides and synaptic vesicle proteins such as synaptic vesicle protein and synaptotagmin. Gangliosides bind the specialized SXWY motif present on ganglioside-recognition site in bacterial serotypes but not for the serotypes of C and D [594–598]. BoNTs for the protein receptor or ganglioside receptor are BoNT/B for the synaptotagmin II, BoNT/A for the GT1b, and BoNT/F for the GD1a [598]. The mechanism for the BoNT-protein receptors or ganglioside receptors is not understood yet. The complexed structure of the BoNT-protein receptor or BoNT-F-ganglioside GD1a has been elucidated [588]. The GD1a-BoNT interaction does not affect the Syt-II toxicity, giving the independent binding. In the BoNT/B-ganglioside interaction, the ganglioside-recognizing site on BoNT/B is a conserved SXWY motif [596]. GD1a, having a Neu5Ac α2,3βGal1,3βGalNAc-glycan structure, recognizes the ganglioside-recognition site of BoNT/B, in a mode that a SA generates hydrogen bonds linked to four amino acid residues of N1273, N1105, G1277, and Y1263. Another SA yields a hydrogen bond to amino acids E1190 and W1262. The structures of BoNT/A-GT1b interaction [599] and BoNT/B-sialyllactose interaction are also reported, where GT1b and GD1a bind to BoNT/B. K1265 and R1269, near W1262, bind to the SA moiety in GT1b, not GD1a, allowing higher affinity of GT1b than GD1a [600]. By analytic tools including computer simulation, NMR spectroscopy, and X-ray crystallography, the glycan minimal binding motif has been suggested. The trisaccharide epitope in the terminal branch of GD1a has been recognized as a minimal binding sequence, and this motif corresponds to the sialyl-Thomsen Friedenreich carbohydrate from the carbohydrate ligand-toxin complex.

Tetanus is a muscle spasm-causing infection, also termed trismus or lockjaw. The bacterium *Clostridium tetani* causes tetanus paralysis because it is an extremely powerful neurotoxin with an approximately calculated human lethal dose of 0.5 pg/kg. Several toxins such as tetanospasmin and tetanolysin are known for the symptom of tetanic spasm. Among them, tetanospasmin influences the nerve-muscle motor interaction, which causes rigidity and muscle spasms but tetanolysin causes tissue damages. By the retrograde axonal transport, tetanus neurotoxin (TeNT) reaches neurons and consequently blocks neurotransmitter release. The TeNT targets, attaches, and enters the terminal presynaptic sites of the motor neuronal neuromuscular plate, triggering blocking of GABA and glycine releases. The results are expressed as paralysis in muscle fibers. The blocked releases of GABA and glycine as the inhibitory neurotransmitters cause failure in inhibition of the motor neuronal responses to sensor responses. This symptom is called tetanic spasm.

Tetanus immunoglobulin prevents or treats the disease together with antibiotics and neuromuscular blockade. Excretion of inhibitory neurotransmitters is strictly inhibited by TeNT at the spinal cord, causing a paralysis [601]. The TeNT protein has a molecular weight of 150 kDa with three domains. Each domain is a MW 50 kDa. TeNT is composed of an L (light) chain with the molecular weight 50 kDa and an H (heavy) chain with the molecular weight 50 kDa. The subunits are S-S bonded and the TeNT toxicity is attributed to the L-chain, a Zn^{2+} metalloprotease to degrade the synaptobrevin 2 and SNARE [602], where the synaptobrevin is known as a key molecule for the neurotransmitter excretion in synapses. The cleavage of SNARE protein elicits the blockade or release of the inhibitory neurotransmitters. The C-terminal H-chain (HC) binds to host cells to internalize into vesicles [603]. Two domains of the H-chain bind to neuronal cells and result in endocytosis. The N-terminal H domain functions to translocate the L-chain across the vesicle membrane to the neuronal cytosol [604]. For the TeNT of *C. tetani*, it exhibits the similar toxicity as BoNTs. TeNT is also the causative agent of the neuroparalytic disorder, inhibiting neurotransmitter release. But, the molecular explanation on the TeNT translocation via endosomal membranes is not understood. Its endocytosis to neurons is initiated by interaction with gangliosides and then further binding to a protein co-receptor.

The TeNT recognizes gangliosides on cell surfaces and a protein receptor. From the study on HC-ganglioside complex, HC has been known to have two ganglioside-binding sites of the Gal4-GalNAc3 and SA-binding sites. HC mutation diminishes the binding capacity to the gangliosides on neuronal cells. TeNT recognition of the carbohydrate motif of gangliosides involves Sia residue. The SA-binding pocket contributes to endocytosis of the TeNT by hydrophobic interactions between the SA-binding site and carbohydrate. The ganglioside b-series of GT1b and GD1b are used not only for the TeNT receptor of host cells, but also neuronal membranes are much better receptors for the TeNT than such gangliosides [604]. As the di- and tri-sialogalactosyl biantennary types, GD1b and GT1b gangliosides are present at high density on neuronal cells and tissues. The GD1b or GT1b binding to TeNT is mediated by the H-chain. The H-chain with free sulfhydryls strongly recognizes the GT1b and the binding capacity is diminished by the disulfide S-S bond formation. TeNT recognizes gangliosides via an additional hydrogen bond formation between the GalNAcβ1,3 carbohydrate residue and amino acid Asn-1219 [510, 605]. TeNT-ganglioside binding to the core GalNAcβ1,3Galβ1,4 sequence is not affected by the additional SA (NeuAc) residues [606].

5.8 GM1a/GM2-*Clostridium perfringens* Alpha-Toxin and Delta-Toxin

Clostridium perfringens is known to produce various extracellular toxins and causes intestinal diseases in humans. Individual toxins are typically classified into five toxin types A to E by Alpha, Beta, Epsilon, and Iota forms. For alpha form, GM1a is the primary cellular receptor and thus often regarded as a strategic target for therapeutic drug design to the strain. GM1 recognizes TrkA and modulates by directly interacting with TrkA in the PM lipid rafts [607]. Asialo-GM1 does not bind to alpha form. GM1a protects membranes fluidity from the alpha form-disrupted environment; however asialo-GM1a does not [608]. All the three gangliosides of GD1b, GQ1b, and GT1b contain the same glycan structures as GM1a glycan. However, the only difference between them is in SA lengths. They bind to alpha form toxin, but less efficiently bind to the toxin than GM1a binding, indicating that the GM1a-like structure is required. Alpha form toxin-GM2 binding with the same SA carbohydrates as GM1a is less strong than GM1a. Thus, Galβ1 at the GM1a binds to alpha form toxin [609]. Alpha form toxin has also high affinity for b-series GT1b and GD1a compared to a-series GM1a and GD1a gangliosides [608].

The anaerobic growing *C. perfringens* [610] induces myonecrosis or gas gangrene in humans [577, 578]. *C. perfringens* alpha form toxin elicits superoxide, chemokines, and cytokine releases by the TrkA-GM1a complex. Toxin α-form is a 43-kDa protein with N-domain for the catalytic site and C-domain for membrane-binding site. Toxin α-form has a phospholipase C (PLC) enzyme activity important for downstream signaling pathway. DAG accumulated at the PM in α-toxin-added cells induces GM1a-associated clustering process and phosphorylation of TrkA. The N-terminal loop region at amino acid 72–93 residues contains a ganglioside carbohydrate-recognition motif (H-SXWY--G) like botulinum toxin sequence. It recognizes the GM1a carbohydrate structure of Neu5Acα2,3(Galβ1,3GalNAcβ1,4) Galβ1,4Glcβ1Cer [579]. Using confocal microscopy with labeled GM1a, the toxin has been confirmed to colocalize with GM1a. The alpha-toxin is less toxic in β1,4-N-GalNAc-transferase KO mice, lacking GM1a. However, this toxin is more toxic in α2,8-sialyltransferase KO mice, due to deficiency of other two b-series ganglioside and c-series ganglioside. Thus, GM1a interaction with the alpha-toxin activates the TrkA receptor signaling [579]. GM1a in membranes promotes TrkA enrichment in lipid rafts with downstream signal [580]. Then, DAG induces interaction of GM1a and TrkA. Thus, GM1a is used for a potential target development against such pathogens.

Delta toxin belongs to a hemolysin generated by C strain and B strain of *C. perfringens* [581]. The Delta toxin protein is a basic protein with an alkaliphilic pI 9.1 and MW 42 kDa protein that hemolyzes erythrocytes in sheep, goats, and pigs [582]. Delta toxin is broad in its cytotoxicity to macrophages, monocytes, and blood platelets [581, 583]. The receptor of Delta toxin is the ganglioside GM2 [544]. Delta toxin specifically binds to GM2 in membrane [584]. Thus, Delta toxin can be used for the probing tool for GM2 existence. Delta toxins have different pore-forming activity to erythrocytes and leukocytes, recognizing ganglioside GM2 on the target cell surface [544, 585]. For example, a type Delta toxin destroys sheep erythrocytes and HeLa cells if GM2 is shed in the PM. Delta toxin selectively can agglutinize and lyse GM2-carrying malignant tumor cells, as Me180, A375, and C1300 tumor cells can be easily targeted. In addition, in vivo treatment of Delta toxin to such tumor-bearing mice eliminates tumor behavioral properties [585]. There is also a possibility that Delta toxins use a co-receptor with GM2 and another membrane protein because Beta toxin does not destroy sheep erythrocytes or HeLa cells. In addition, the beta toxin does not bind either to GM2 or GMI and it does not recognize HeLa cells. Beta and Delta toxins possibly co-utilize the pore formation mechanisms from the homology similarity, but the two toxins recognize each specifically different receptor on target cells. The receptor binding motif is present on the carboxy terminal region of Delta toxin and Delta122–318 binds to the target HeLa cells. Sequence homology analysis of the carboxy terminal region of Beta and Delta toxins exhibits the carbohydrate receptor-binding site.

5.9 GD1a-*Bordetella pertussis* Toxin

Bordetella pertussis produces exotoxins of filamentous hemagglutinin and pertussis toxin (PTx). *B. pertussis* adhesin PTx binds to glycans expressed on human cilia and macrophages [586]. The B oligomer-GD1a binding is mediated by SA residues for the *B. pertussis* toxin [542]. Vaccination using PTx, a main virulence factor of *B. pertussis*, has been attenuated in its role, reducing whooping cough (named pertussis) mortality [587]. As a non-cellular pertussis vaccine, PTx is a vaccine candidate for innate and adaptive immune immunity [224, 588]. Pertussis, called whooping coughs, is a respiratory disorder and its causative agent is the etiological *B. pertussis* strain of Gram-negative form. PTx exhibits systemic lymphocytosis and hyperinsulinemia. PTx belongs to a family of the AB5 group like *V. cholerae* CTx, *E. coli* LT, *E. coli* STx, and *S. dysenteriae* STx. The AB5 toxin forms hexamer protein complexes with five ligand-recognition subunits (B) and a monomeric enzymatic subunit (A) as the ring structure. The PTx is the AB5 toxin and hexamer protein complex of 5 B subunits and one A subunit with an enzymatic property. Pertussis toxin consists of an enzymatic component A protomer, PTx A (S1, Mr-28,000) noncovalently bound to the PTx B. PTx B is composed of four dissimilar subunits of S2 (Mr 21,900), S3 (Mr 21,900), S4 (Mr 12,000), and S5 (Mr 11,750) in a mode of 1:1:2:1 molecular ratio. Therefore, PTx is also a Trojan horse type. The PTx A subunit, termed S1, is an ADP-ribosyltransferase, an enzyme specifically ribosylates the α-subunit of several GTP-binding proteins in cell surfaces. Enzymatically, PTx catalyzes ADP-ribose transferring reaction from substrate NAD+ by S1 to the acceptor α-subunit (Gα) of GH protein and consequently the ADP ribosylation leads to the prevention of Gα action with adenylate cyclase inhibition. Therefore, it is a specific ADP-ribosylation enzyme, which is the α-subunit of GTP-binding proteins, preventing coupling of the G proteins from the receptors. Unlike classical AB5 toxins with 5 B subunits as an identical form, the pentameric PTx B has the four different subunits of S2, S3, S4, and S5, as described [589]. However, S3 subunit recognizes glycosylated and phospholipid receptors on the host cell PM. The A subunit S1, which is the ADP-ribosyltransferase enzyme, targets the GTP-binding α-subunit protein [590]. The B-pentamer targets cells and S1 entry into host cells. B antigen activates antigen-independent T cells and mitogenicity [589, 591].

PTx binds primarily to the carbohydrates surfaced on cell receptors and has no affinity for the polypeptides. The pentameric PTx B is necessary for cell recognition, attachment, and entry of S1 to target cells. PTx B activates antigen-independent T cells. This receptor-binding PTx B is the new concept in the AB toxin biology. PTx binds to the glycan residues, not proteins. GD1a (MW, 1836 Da), a disialoganglioside, is a receptor specific for the pertussis toxin PTx B. Several surfaced proteins including glycoprotein Ib, CD14, TLR-4, and Mac-1 (CD11b/CD18) are the known PTx receptors [592–594]. Among several known PTx receptors such as CD14, glycoprotein Ib, Mac-1, and TLR-4, only glycoprotein Ib exhibits a direct binding capacity to PTx B. In other candidates, serum glycoproteins of fetuin and haptoglobin can be recognized by PTx B and therefore, the PTx-fetuin binding has been investigated. In fact, serum PTxs easily binds the fetuin and haptoglobin. Such binding property of PTx gave the toxin purification strategy using the binding proteins [595]. Fetuin is terminally glycosylated with Neu5Acα2,3Gal and Neu5Acα2,6Gal tri-antennary N-glycans. PTx recognizes, in a high affinity mode, Neu5Acα2,6Gal rather than Neu5Acα2,3Gal. In addition, elimination of the terminal Neu5Ac residue, conversion of the NeuAc to NeuGc form, or terminal fucosylation on the core structure of N-glycan carbohydrates reduces the PTx binding capacity to the host cells [596]. PTx targets terminal Neu5Ac. NeuGc fucosylation abolishes the binding capacity [597]. The PTx B parts of S2 and S3 exhibit 77% identity and consist of two SA-binding sites, named 2SA and 3SA, each corresponded to the amino acid position of 101–105 and 125. Among the SA and Gal residues in the PTx B-glycan complex structure, only the terminal SA residues directly recognize the PTx B protein. The suggested SA-binding PTx B S2 site has a neutral glycolipid-binding sequence at the amino acid sequences of 37–51, a WGA agglutinin homologous sequence at the amino acid sequences of 18–23, and aerolysin protein-homologous sequences at amino acid sequences of 82 and 52–72. PTx B S3 site has aerolysin-homologous sites at the amino acid sequence of 82 and 52–72 and a ganglioside-recognition site at the amino acid residues 37–51 sequences [598].

PTx has a lectin function with considerable plasticity. Four glycan binding sites are currently suggested. Two SA-recognition sites present in the S2 and S3 proteins in the C-terminal regions and also asialoglycan-carrying N-glycans of the N-terminal S2 protein and S3 protein are known for the four glycan-binding sites. Rather than the N-terminal regions of S2 and S3 sites, the S2 and S3 sites in the C-terminal region are specific for only a SA residue. ST enzyme ST6Gal-I catalyzes SAα2,6 residue transfer to N-glycans and ST enzyme ST8Gal-I catalyzes SAα2,8 residue to gangliosides in the brain. ST enzyme ST8Gal-IV is specific for the SAα2,8 residue transfer to N-glycans. Although some plant-producing lectins and AB5 type toxins have multiple binding sites, PTx recognizes multiple glycan targets [599]. For regulation of the ADP-ribosylation, PT ADP-ribosylation activity is inhibited by GQ1b, because GQ1b inhibits the C3 exoenzyme, demonstrating the inhibitory effect of gangliosides on ADP-ribosylation. Therefore, gangliosides involve in the cellular NAD+ metabolism [600]. Similar to PTx, it is known that the C3 exoenzyme generated by *C. botulinum* C and D strains catalyzes the Rho family ADP-ribosylation, where Rho protein is known as a small-sized GTP protein of the host cells. ADP-ribosylation catalyzed by C3 exoenzyme interrupts the Rho protein interactions.

5.10 GM1/GM/Complex Ganglioside-*Helicobacter pylori*

Helicobacter pylori recognizes gangliosides and has long been known as a primate and human infectious bacteria in the mucosal stomach layer or the gastric epithelium. The bacterial genome size is smaller than that of *E. coli*, by one-third size of *E. coli* [601]. *H. pylori* adhesion to hyperglycosyl mucins on the mucus layer of epithelium potentiates bacterial colonization without disturbed infection caused by apical turnover and mucus layer shedding in epithelium [602]. *H. pylori* is regarded to be adapted to the gastric environment through binding to cell surface carbohydrates, named receptors or binding epitopes in the mucosa.

5.10.1 *Ganglioside Binding to* H. pylori *Vacuolating Cytotoxin (VacA) Prevents its Toxin Activity*

Gangliosides directly bind to *H. pylori* surface vacuolating cytotoxin (VacA) and consequently neutralizes the toxin activity. Gangliosides inhibit VacA-mediated vacuolation in cells such as AZ-521 cells. Lyso-GM1, lyso-GM2, or lyso-GM3 gangliosides inhibit VacA-formed vacuolation in host cells such as AZ-521 cells, when VacA internalizes to AZ-521 cells [603]. Binding VacA or cholera toxin B, CTB, to lyso-GM1-conjugated resin such as Sepharose has been confirmed. The lyso-GM1-binding protein has been traced in *H. pylori*. Lyso-GM1, lyso-GM2, and lyso-GM3 binding to VacA has been examined using FCS. GM1 shows higher binding affinity to VacA rather than GM2 and GM3. Examined results for the neutralization capacity of the GM3 on vacuole formation in the host cells suggested that gangliosides bind to VacA and neutralize its toxicity. Fatty acids of ganglioside are not required for its binding to VacA. Both sphingosine and carbohydrate chain are necessary for binding to VacA. The carbohydrate moiety 3'SL of GM3 inhibits vacuole formation in AZ-521 cells. The vacuole is also formed by *H. pylori* VacA, where GM1 binds directly to VacA [604]. *H. pylori* VacA recognition with GM1 neutralizes the VacA toxin activity. *H. pylori* can adhere and bind to erythrocytes and neutrophils due to their mucosal carbohydrates [605]. These carbohydrate glycans are presented on glycoproteins such as gastric mucins such as MUC5AC and MUC6 and GSLs [611]. The carbohydrates include fucosyl ABO antigens [606], sialylated carbohydrates [607], sulfated carbohydrate [608], and core glycans unsubstituted [609]. *H. pylori* synthesizes many attachment proteins, adhesin [610]. The well-known recognition event of the *H. pylori* adhesins and receptors present on host cells is based on the molecular recognition of blood group-binding adhesin (BabA) and ABO antigen as well as BabA and Le^b antigen [606]. The Le^b and ABO group carbohydrates are found in erythrocytes as antigens of histo-blood group glycan, while the histo group glycan antigenic epitopes are similarly present in the epithelial tissues and gastrointestinal tract (GIT) mucus [612].

The *H. pylori* binds to complex gangliosides, as has been demonstrated and characterized by MS and proton NMR [613]. *H. pylori* strains recognize SA-containing glycoconjugates [609, 613–623]. Some receptors are gangliotetraosyl-Cer, the Le^b antigen, NeuAcα2,3-neolactotetraosyl-Cer, Lac-Cer, and Lactetraosyl-Cer [614, 618, 620]. The sialyl-dimeric-Le^X of NeuAcα3Galβ4(Fucα3)GlcNAcβ3Galβ4(Fucα3)GlcNAcβ3Galβ4Glcβ1-Cer [619] and gangliosides sialyl-neolactohexosyl-Cer of NeuAcα3Galβ4GlcNAcβ3Galβ4GlcNAcβ3Galβ4Glcβ1-Cer [619] are also known for *H. pylori* receptors. Representatively, the *H. pylori* receptors indicated GM3 and sulfatide [623].

With regard to the gastric epithelium, different carbohydrate receptors are recognized by *H. pylori* [624]. However, three carbohydrate glycans binding adhesins are known. They include BabA [625] and SA-binding adhesin, SabA as the adhesin molecule [607] as well as the LacdiNAc-binding adhesin (LabA) [613]. BabA binds to Le^b and fucosyl antigens as the basic adhesin to the gastric mucosal region of humans. SA-binding adhesin (SabA) binds to LacNAc-containing gangliosides with terminally attached NeuAcα2,3-linkage, with longer LAcNAc chains, and binds to Fuc-substituted LacNAc core chains such as dimeric sialyl-Le^{X} [626]. LacNAc or CD75 has a structure of Galβ1,4GlcNAcβ1-R and LacNAc unit is often present on terminal residue of GSLs and glycoproteins. CD75s (α2,6-sialyl LacNAc, Neu5Ac α2,6Galβ1,4GlcNAc β1-R) is a form of α 2,6-Sialylation of CD75 that forms the CD75s epitope [α2,6-sialic acid involves the exocyclic hydroxymethyl group, in contrast to the a2,3-connection to a C-atom in the ring in CD15s or CD65s] [627]. This special extension choses each tissue lectin of galectins or siglecs. Histo-blood group H type 2, CD175, has a carbohydrate structure of Fucα1,2Gal β1,4GlcNAcβ1-R. Similar to CD75s (75 s) having the LacNAc core (CD75), but not to a2,6-sialylation (CD75s), CD173 contains LacNAc having a1,2-fucosylation (by the fucosyltransferase FUT1) on glycoproteins and glycolipids and forms the histo-blood group H (type 2) structure. In non-Le^{b} cases, *H. pylori* is colonized by SabA adhesin and sialylglycans [627]. The human stomach acid GSLs include sulfatide as well as GM3, GM1, GD3, and GD1a. Two SabA-recognizing ganglioside structures of *H. pylori* strains are Neu5Acα3-neolactohexaosyl-Cer and Neu5Acα3-neolactooctaosyl-Cer, whereas human stomach acidic GSLs including sulfatide, GM3, GD1a, GD1b, GD3, and Neu5Acα3-neolactotetraosyl-Cer are not bound by the SabA protein of *H. pylori* [628].

The *H. pylori* adhesin, SabA, binds sialyl-Le^{X}. *H. pylori* J99 and CCUG 17874 recognize SA residues in gangliosides through the NeuAcα2,3Galβ1,4GlcNAcβ1,3Galβ1,4GlcNAcβ-recognizing neutrophil-activating protein (HPNAP) or the SA-specific SabA molecule [559, 629]. The bacterial SabA recognizes the LacNAc-carrying gangliosides with terminal SAα2,3-linkage. However, other related structures such as NeuGc-gangliosides and NeuAcα2,6 or NeuAcα2,8NeuAcα2,3 are not recognized [617, 618]. Recognition of *H. pylori* with NeuAcα1,3-neolactooctaosyl-Cer over NeuAcα3-neolactohexaosyl-Cer and NeuAcα3-neolactotetraosyl-Cer has been suggested. This effect has produced increased accessible level of the carbohydrate head group. The binding affinity is increased by numbers of the LacNAc carbohydrate chain lengths, the carbohydrate branches, and Fuc substituted LacNAc core chain site. The SabA adhesin of *H. pylori* is so far the only factor bound to gangliosides. HPNAP is an immunogenic agent of *H. pylori* cell surface [623]. Therefore, bacterial cells bind to NeuAcα2,3Galβ1,4GlcNAcβ1,3Galβ1,4GlcNAcβ-terminated glycosphingolipids. However, HPNAP gene disruption still allows its recognition of GSLs with terminal NeuAcα2,3Galβ1,4GlcNAcβ1,3Galβ1,4GlcNAcβ. In contrast, the recognition of NeuAcα2,3Galβ1,4GlcNAcβ1,3Galβ1,4GlcNAc-β-terminated gangliosides and other gangliosides has been diminished by the SabA adhesin gene disruption. Therefore, it is indicated that *H. pylori* recognizes the ganglioside solely by the SabA adhesin [613].

5.11 Interaction between GalNAcβ1,4Gal in Asialo-GM1 and Asialo-GM2 with Miscellaneous Pathogens of *Legionella pneumophila, Pseudomonas aeruginosa, Burkholderia pseudomallei, Chlamydia pneumoniae, Moraxella catarrhalis, Bacillus anthracis, Francisella tularensis, Burkholderia pseudomallei, Brucella abortus, Yersinia pestis, and Staphylococcus aureus* Enterotoxin B

Viruses and bacteria-causing diseases can be treated with antiviral and antibiotic agents, respectively, but exotoxins-caused diseases can be treated with toxin-reactive antisera. However, the current issues of antibiotics resistance or toxicities of antiviral agents are emerged to develop alternative agents. From the fact that adhesins interact with carbohydrates, the prospective concept of the anti-adhesion molecules has been emerged as a new therapeutic strategy to alternate conventional drugs. For example, carbohydrate-recognizing property of ricin, a plant-derived toxigenic lectin, is clarified and the Galβ1,4GlcNAc structured candidate has been suggested as its anti-adhesion drug. In respiratory bacterial endospores such as *Bacillus anthracis* [630], it has been known that their produced adhesions bind to the GalNAcβ1,4Gal sequence on ceramide structures as the common disaccharide structure from the asialo-GM1 and asialo-GM2. *B. anthracis* as the archetypal agent produces endospores. The endospores attach and internalize into macrophages. After infection, the vegetative cells generate adhesive molecules, termed adhesins, which include many different molecules such as S-layer proteins or collagen–/fibronectin-binding proteins [631]. Several carbohydrates, which were mimicry to the binding sugars, of GalNacβ1-3Gal, dextran sulfate, Galβ1-4GlcNAc, and GalNAcβ1-4Gal block the adhesin binding and internalization to the macrophages [632]. Pathogens and toxins multivalently recognize glycans or glycan receptors present in the cell surfaces of hosts (Fig. 5.1). The carbohydrate receptors consist of oligosaccharides structures that are structurally diverse in their features including linkages and branching

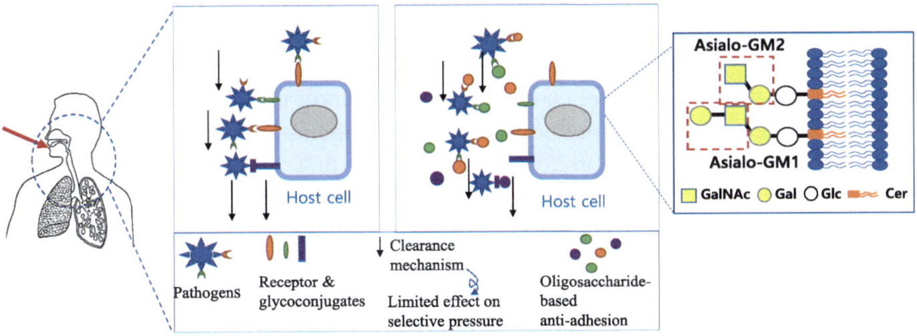

Fig. 5.1 Bacterial carbohydrate-mediated attachment to host cells

patterns of sugar units on proteins or lipids. The levels of ganglioside GM1, asialo-GM1, GM2, and asialo-GM2 are varied in classes depending on the tissue-specific epithelial cells in respiratory track system. Many respiratory pathogens attach and recognize host carbohydrates (Table 5.1). For example, the GalNAcβ1,4Gal disaccharide structure is recognized as a minimal length of carbohydrate moieties. This sequence is necessary for pathogenic adhesion for the respiratory infectious agents and is present in the asialo-GM2 and asialo-GM1 gangliosides [630, 633].

Many respiratory pathogens are known to bind to specific carbohydrate sequences [630] (Table 5.1). Pathogens also generate carbohydrate structures in the capsules, glycoproteins, and LPS. They directly attach to the cell surfaces of host epithelium or innate immune cells. The minimal glycan unit is the disaccharide GalNAcβ1-4Gal that is important for the attachment of asialo-GM1 and asialo-GM2 [634–639]. Pathogens produce carbohydrate structures as capsules, LPS, and glycoproteins to attach the host and immune cell surfaces. For the anti-adhesion therapy, anti-adhesion oligosaccharides such as sialyllactose prevent *H. pylori* infection. *B. anthracis* endospores interact between the exterior endospore and human tissues, although the mechanisms of attachment are not well understood for vegetative bacterial cells and bacterial endospores. The bacterial endospores contain an exosporium form with filamentous appendages and glycoproteins for endocytosis into innate immune cells such as macrophages [640, 641]. During germination, the vegetative cell adhesins attach to host cells [642]. Glycans such as dextran sulfate, GalNAcβ1,4Gal, GalNAcβ1,3Gal, and Galβ1,4GlcNAc inhibit attachment of bacterial vegetative cells to the host cells [643].

Legionella pneumophila induces Legionnaires' disease in alveolar macrophages. *L. pneumophila* can grow within monocytes and macrophages as well as within amoebae that is probably the natural host [635]. Intracellular replication is a key virulence determinant. Macropinosomes are enriched with GM1 and glycosylphosphatidylinositol-linked proteins. *P. aeruginosa* is a major death causative agent with cystic fibrosis and the bacterial receptors are not known. The asialo-GM1 (Gg4Cer) and asialo-GM2 (Gg3Cer) are known as specific binding receptors for the bacteria in solid-phase binding assays. As the Pseudomonas is negative for binding to LacCer, the β-N-GalNAc residue in internal asialo GM1 and terminal asialo GM2 is suggested for recognition. The glycan β-N-GalNAc itself seems not to bind, because the bacteria do not recognize globosides or Forssman glycans. *P. aeruginosa* and *P. cepacia* bind to GalNAc β1-4Gal residues in GSLs as the bacterial receptors [636].

Burkholderia pseudomallei causes melioidosis upon infection, because the bacteria are the life-threatening pulmonary infectious agents in tropical regions. Adhesion and attachment to respiratory cells are the first step of host infection of *B. pseudomallei*. *B. pseudomallei* causes melioidosis symptoms which encompass pulmonary infection. *B. pseudomallei* attaches to epithelial cells such as alveolar tissue, bronchial tissue, cervical tissue, conjunctival tissue, laryngeal tissue, nasal tissue, and oral tissues-originated cells by asialo-GM1 and asialo-GM2 [637]. *B. pseudomallei* acid phosphatase also recognizes asialo-GM1 and asialo-GM2. The carbohydrate sequences of the GalNAcβ1,3Gal-structure and the

Table 5.1 GSL-based receptors for exotoxin forms of microbial virulence factors and receptor roles

GSLs	Toxin or infectious agents	References
GM1	Cholera toxin	Jobling MG et al. 2012 [286], Wernick NL et al. 2010 [287]
	monkey polyomavirus SV40 virus	Neu U, 2008 [329-1]
	Bovine rotavirus (UK strain)	Delorme C et al., 2001 [41]
GM3	Bovine rotavirus (NCVD strain)	Delorme C et al., 2001 [41]
GM1, GD1b, SAα2,6-Lac series tetrasaccharide	human polyomavirus JC (JCPV)	Neu et al 2010 [336]
GD1b, GT1b, SAα2,3-R	human polyomavirus JC (BKV)	Low JA et al. [2006] [51], Dugan AS et al [2005] [338]
GD1a, GT1b, GM1, SAα2,3-R	mouse polyomavirus (mPyV)	Tsai B et al. (2003) [53]
GT1b, SAα2,3- or SAα2,6-R	JCV	Dugan AS et al. [2008], Komagome R et al. [2002]
GD1a, GT1b, GQ1b, neoLac-series gangliosides of SAα2,3-R	paramyxoviruses (Respirovirus, Rubulavirus, Avulavirus)	Villar E, Barroso IM. (2006) [291], Anastasia L, et al (2008) [352]
GD1a	Adenoviruse type 37 (Ad37)	Nilsson EC et al. 2011 [234]
Gb3	Shiga toxin	Karve SS et al., 2014 [33], Lingwood CA. 2996 [93], Saxena SK et al. 1989 [269-1], Jacewicz M et al. 1986 [167]
	Verotoxin	Surendran-Nair M et al., 2016 [34], Lingwood CA et al. 2010 [99]
GalCer	HIV	Bhat S et al. 1991 [199-1], Bhat S et al., 1993[35], Hug P et al., 2000 [39]
GM3, Gb3	HIV	Hammache D et al., 1998 [38], 1999 [37],
Gb3	HIV	Lund N et al., (2006) [40], Rawat SS et al. 2005 [40-2], Lingwood CA et al. 2010 [99]
GD3	HIV	Andersson LM et al. (1998) [40-1], Hammache D et al. 1999. [37].
Gb4	Parvovirus B19 (B19V)	Rogo LD et al. (2014) [257-1], Brown KE et al. 1993 [257-3], Bönsch C et al. 2010 [257-7]
GT1b, GQ1b, GD1a, GD1b	botulinum toxin	Benson et al. 2011 [586], Kozaki et al. 1998 [587]
GT1b, GD1b	tetanus neurotoxin	Halpern e al. 1995 [591], Burns et al. 2017 [594]
GD1a	pertussis toxin	Millen et al. 2010 [585]
LacCer, globoside, asialo-GM1, asialo-GM2	Yersinia pestis adhesin	[637,638]
DiGalCer	Staphylococcal enterotoxin B	[645]
Asialo-GM1, LacCer	Ricin	[645-1.645-2]

GalNAcβ1,4Gal-structure appeared in the asialo-GM1 ganglioside and asialo-GM2 structures inhibit attachment and interaction of *B. pseudomallei* to the alveolar pneumocytes of human tissues. The asialoganglioside GM1 and asialoganglioside GM2 inhibit the attachment of *B. pseudomallei* upon treatment, indicating that asialogangliosides GM1 and GM2 are the binding receptors in epithelial cells. *Chlamydia* species including *C. pneumoniae* are Gram-negative and obligate intracellular bacteria, causing pneumonia, sinusitis, and bronchitis. The chlamydiae of *C. pneumoniae*, *C. trachomatis*, and *C. pneumoniae* also recognize asialo-GM1 and asialo-GM2. However, they do not bind to neutral or acidic GSPs [638]. *Moraxella catarrhalis* is a respiratory and middle ear infectious pathogen. It attaches to pharyngeal epithelial cells though GM2 by adhesins or receptors. *M. catarrhalis* recognizes asialo-GM1 and asialo-GM2. GalNAcβ1,4Galbeta1 sequence found in Gg4Cer and Gg3Cer acts as the receptors for host cell attachment [639]. The genus *Brucella* causes brucellosis as an infection disease in mammalian species. Brucella species infects many animal species such as sheep, pigs, and humans, producing a pneumonia. Natural Brucella species are *Brucella ovis* and *Brucella canis*. Brucella infection in macrophages involves cholesterol and ganglioside GM1 associated with lipid rafts [644]. *B. melitensis* and *B. abortus* recognize SA residues on epithelial cells, erythrocytes, and macrophages by adhesins, SP29 and SP41 [644, 645]. NeuAc, N-acetylneuraminyl Lac, and the glycosaminoglycan (GAG) such as chondroitin sulfate block attachment to erythrocytes [646]. *B. pseudomallei* is closely related to *P. aeruginosa* and *B. cepacia*. It frequently exhibits anti-antibiotics spectrum due to multidrug efflux pumps and enzyme mutant. It causes melioidosis with pulmonary infection and wide tissue tropism to epithelial cells from alveolar tissues, bronchial tissues, cervical tissues, conjunctival tissues, laryngeal tissues, nasal tissues, oral tissues and conjunctival tissues [647]. *B. pseudomallei* binds to human epithelial cells via asialo-GM2 and asialo-GM1 [648]. Furthermore, the bacterial phosphatase binds to asialo-GM2 and asialo-GM1 as well as GalNAcβ1,4Gal and GalNAcβ1,3Gal located on asialo-GM2 and asialo-GM1 [632, 649]. The carbohydrate sequences of asialo-GM2 and asialo-GM1 are GalNAcβ1,4Galβ1,4Glc-Cer and Galβ1,3GalNAcβ1,4Galβ1,4Glc-Cer having that the monosaccharide includes Glc, Gal, and GalNAc.

Yersinia pestis as the etiological agent is known as bubonic plague with fatal disease for mortality 100% within 48 h [647]. *Y. pestis* attaches to oligosaccharides on the host cells [648]. Adhesins of Ail, YadA, invasin, and pH 6 antigen (Psa) are known for the yersiniae. Disaccharide structures of GalNAcβ1,3Gal and GalNAcβ1,4Gal carbohydrates are the specific interacting carbohydrates [649]. The GalNAcβ1,3Gal and GalNAcβ1,4Gal glycans found in asialo-GM2 and asialo-GM1 [650] bind to *Y. pestis* PsaA. The PsaA resembles to a fimbrium structure and binds to host cell phosphatidylcholine, the apolipoprotein B lipids and IgG Fc receptors [651]. *Y. pestis* also binds to LacCer and globoside as well as collagen type IV [652]. Pla binds specifically to Gal residues [652]. The Psa antigen requires the β1Gal-residues in GSLs as the minimum determinant. ***Y. pestis*** is well known as the etiological bacterium, causing bubonic plague and pneumonia. Many types of adhesins are known but their ligands are not known. Disaccharides of

GalNAcβ1,4Gal and GalNAcβ1-3Gal have been known for the inhibitors of *Y. pestis* attachment to host cells. The GalNAcβ1,3Gal and GalNAcβ1,4Gal structures as the common carbohydrates of asialo-GM2 and asialo-GM1 bind to PsaA of *Y. pestis*. As adhesins, *Y. pestis* PsaA homolog HecA is also known. Another adhesin of the tad gene cluster known in *Actinobacillus actinomycetemcomitans* and *Y. pestis* nonspecifically attach to the host cells [653]. The YapC and YapE autotransporters are also adhesins. The YapE colonizes lymph nodes and disseminates the bubonic plague forms. The *Y. pestis* LPS binds to DC-SIGN on the DCs and alveolar macrophages toward delivery to lymph nodes [654]. The *Y. pestis* plasminogen activator (Pla) converts plasminogen into plasmin and disseminates *Y. pestis* to whole body. Pla as an adhesin binds to the **glycolipids, LacCer and globoside**. Pla binds specifically to Gal residue of host cells. As described previously, *Y. pestis, B. anthracis, Brucella* species, *Burkholderia pseudomallei,* and *Francisella tularensis* also produce such adhesins. Therefore, the experimental results on oligosaccharide inhibition suggest the common structures of GalNAcβ1,4Gal and GalNAcβ1,3Gal as well as the ceramide moieties. In addition, from the botulinum toxin studies, it has been well known that botulinum serotypes A to F recognize certain GSL forms such as GT1b, GQ1b, GD1a, and GD1b. This indicates the possible designation of mimicry of their binding structures. Respiratory epithelial cells express various gangliosides but generate desialylated glycolipids such as asialo-GM1 and glycoproteins during many pathogenic infections such as influenza virus, *P. aeruginosa and S. pneumoniae*. This phenotype change contributes to presentation of attachment receptors [630]. GalNAcβ1,4Gal in asialo-GM2 and asialo-GM1 enhances the attenuating potency for *P. aeruginosa* and F1C-fimbriated *E. coli* [655].

Staphylococcus aureus is one of the most warranted bacterial pathogens. The initial step of the infection involves bacterial attachment and adherence to host tissue or epithelium. *S. aureus* binds to gangliosides and asialo-GM1. Staphylococcal enterotoxin B (SEB) induces diarrhea. Neutral GSLs of human kidney proximal tubule cells inhibit SEB and the SEB receptor is di-GalCer [656]. For binding targets for toxins, toxins such as botulinum toxin (BoTx), SEB, and ricin recognize carbohydrates. Ricin binds to asialo-GM1 or Lac-Cer [657, 658]. *C. botulinum* BotTx blocks release of neurotransmitters and blocks neuronal potentials. Seven BotTx A–G types are known and recognize gangliosides, synaptotagmins, and phospholipid receptors present on the presynaptic membranes. GT1b inhibits the BoTx activity. BoTx type A recognizes predominantly GT1b and GQ1b as well as GD1a but not GM1 ganglioside.

F. tularensis as a pneumonic pathogen generates cell-surface adhesins called the type IV pili, which adhere to macrophage cells, pneumocyte cells, and hepatocytes. FsaP is an adhesive protein present in cell surfaces of the type II pneumocyte A549 cells. The *F. tularensis* subunit PilA restores type IV pili when expressed in a *N. gonorrhoeae* deficient strain for the type IV pilus [659]. In addition, *F. tularensis* attachment, even in opsonized and non-opsonized forms, to DCs, macrophages, and monocytes occurs through MBP or surfactant protein A produced on the innate immune cells [660]. Brucella genus bacteria including *B. abortus* and

B. melitensis infect animals such as sheep and pigs as well as humans through contact with such animals. They recognize SA residues appeared on erythrocytes, epithelial cells, and macrophage cells through adhesins of SP29 and SP41 termed UgpB [646]. NeuAc, NeuAc-Lac, and the GAG of chondroitin sulfate block the bacterial attachment to erythrocytes. Treatment with certain oligosaccharides block bacterial attachment of *H. pylori, L. pneumophila, S. pneumoniae,* and *P. aeruginosa* [630] Thus, this sequence of the disaccharide unit can be used as a candidate for anti-adhesion agents. Table 5.1 summarizes the glycolipid-based receptors responsible for microbial virulent factors as exo-forms and receptor roles.

Chapter 6
GSL Signaling Regulation

6.1 GSL Signaling Regulation in Neurodevelopment

GSLs as membrane components of cells are most abundant in the CNS, because GSLs confer protection and restoration of neuronal functions. This reflects clinically available avenue for ganglioside replacement therapy for dysregulation of neuronal function. Currently, gangliosides are obtained bovine-free for therapeutics. In the functional side, ganglioside expression profiles are changed during the nervous system development. Neural cell fates are associated with the endogenous GSL synthesis, shifting to complex gangliosides. During NSC differentiation, GD3 is further used for the generation of complex forms including GT1b, GD1b, GM1, and GD1a, and consequently, induces terminal differentiation of the cells and defecting the cell stemness. Neural stem cells (NSC) differentiate into the nervous cells in the brain. Changes in gangliosides expression are a prerequisite in developmental changes of neuronal phenotype of nerve cells. The ganglioside roles and functions are important in NSC function and terminally differentiated neuronal cells. The surfaced composition, distribution, diversity, and localization of globo series and ganglio series lead to proper intercellular recognition, binding, interactions, receptor signaling, and phenotype change. Gangliosides as abundantly expressed in neuronal cell membranes function in neuronal and brain development. The neuronal events such as neurotransmission, synapse formation, and neural circuits are prerequisites for memory and learning. To do such mission, gangliosides are abundant in the neuronal tissues and subjected to be changed in the brain drastically during development by stage-specific expression type or epigenetic regulation in neuron development and differentiation of NSCs [661]. The nuclear distribution and ganglioside composition during the development of the brain reflect the phenotype status of the brain [662]. Resultantly, the brain neuroplasticity is derived through synaptic plasticity generated in the nigro-dopaminergic and hippocampus pathways. Neuronal cell gangliosides constitute 75% of the sialylated molecules in the brain [663]. The glycome as the ganglioside profile changes dramatically during neuronal

C.-H. Kim, *Glycosphingolipids Signaling*,
https://doi.org/10.1007/978-981-15-5807-8_6

development as well as life in a region-specific manner. The tight control of ganglioside biosynthesis designates step-by-step brain maturation, because the changes in brain maturation is indicative for neurodevelopmental milestones, as expressed in stage-specific tube formation in neural area, differentiation, axon formation, dendrite outgrowth, and synaptogenesis.

During neural development, the GSL globo-series including SSEA-3 and -4 are produced [664]. During progressive development, the lacto-series GSLs such as lactoneotetraosyl-Cer (nLc4Cer) appear at embryonic day 1.5 (E1.5) and the GSL ganglio-series are appear at E7. The GlcCer formation commences with the early embryonic genesis of the brain and declined at the stage of E16 [665]. Thereafter, the GlcCer is gradually derivatized to a series of GSLs by Gal-transferase and to gangliosides by STs. Apart from gangliosides, ceramide is converted to GalCer series by Gal-transferase III, named Gal-transferase-III or GalCer synthase in the ER. Consequently, unusual GSLs such as GM4, sulfatide, and GalCer are generated [666]. GSLs maintain myelin nerve function. The sulfatides or complex ganglioside deficiency leads to neurodegenerative outcome due to their roles in maintaining myelin and axonal integrity [667]. Neuronal ganglioside b-series of GD1b and GT1b are crucial for maintenance of survival, because sulfatide and b-series ganglioside coordinate axo-glial adhesion and paranodal setting. The changes in the ganglioside expression reflect the functional role of GalNAc-bearing ganglio-series GSLs in neuronal development. Complex gangliosides are generated through a distinct enzyme, β1,4-GalNAc-transferase 1, termed GalNAc-T, in neuronal cells. Sulfatide such as 3-O-sulfo-GalCer is found on the outer side of leaflets in the myelin and neurons [667]. Sulfatide and complex gangliosides maintain and stabilize nervous systems for the axon and glial interactions in the node of Ranvier. GalNAc-T KO mice exhibit neurodegenerative genotype [668, 669] with the decreased MAG level of the myelin. GD1a and GT1b as MAG receptors are abundantly present on the axonal membrane. MAG-KO mice are featured of delayed node maturation, like GalNAc-T KO mice, exhibiting abnormal nerve formation [670]. MAG synthesis is decreased in GalNAc-T KO mice, linking a phenotype relationship of GalNAc-T KO with MAG KO mice. Also, complex gangliosides are linked with MAG toward the axonal-glial interaction. Therefore, inbreeding mice between GalNAc-T KO and MAG KO mice complement but GalNAc-T KO mice and CST KO mice show independent phenotype. MAG is a myelin membrane receptor for axonal gangliosides. GalC and sulfatide stimulate MAG functions and MAG-ganglioside recognition induces axo-glial integrity.

Neuronal progenitor cells and neurons are phenotypically featured with GM1 expression [669]. GM1 in the nuclear membranes promotes neurite outgrowth of primary neurons. NSCs treated with GM1 promote neurogenic activity [671]. The GM1 treatment elicits GM2 and GD2S gene expression in NSCs as well as induces histone H4 acetylation on promoter cis-region of the GM2-GD2 synthase gene. Thus, ganglioside expression is controlled by epigenetic gene regulation of ganglioside synthetic enzyme genes over neuronal cell differentiation and development. GM1 regulates neuronal differentiation of NSCs through brain-specific gangliosides of GT1b/GD1b/GD1a. But, GM1 ganglioside has a functional role to modulate ganglioside biosynthesis during the developing brain. For example, neuronal cells should survive during the embryonic stage because it is reported that approximate half of the growing neuronal and glial cells are subjected to apoptotic cell death during the gestation day of E12 and of E18 [672]. The gestation stage is the same as a biosynthetic shift from GM3/GD3 to the b-series GD1b, GT1b, and GQ1b at day E14 and E16. This similar biosynthetic switch has also been known for neuronal differentiation of teratoma embryonic stem cells [673]. GD3, named CD60a, is involved in the self-renewal event of NSCs, as it is abundant in the embryonic brain, while GT1b/GD1b/GD1a/GM1 is abundantly enriched in the adult brain tissues [646]. GD3 is appeared at the stage of early development and in the neural tube formation. GD3 is predominantly co-expressed and co-stained with SSEA-1 in NSCs, the mouse brain [674], and cultured mouse neurospheres. Secondly, expressed type is GD2 in human NSCs [675]. GD3-expressing cells are also positive for NSC markers including Sox2, Musashi-1, and nestin. Once NSCs are differentiated, the NSCs decline the GD3 expression. Therefore, GD3 is a biomarker for mouse NSCs. GD3 is also expressed in embryonic, postnatal, and adult brain NSCs. GD3-expressing cells but not GD3-deficient cells form neurospheres and differentiate into neuronal cells.

During mice embryo-formation, the simple GM3 and GD3 are greatly shifted to the complex types of gangliosides [665]. In human cortical layers, GD1a level is dramatically increased concomitantly with a progressed cortical synaptogenesis [676]. GM1 and GD1a levels are increased in the frontal cortex with concomitant differentiation of human neurons, dendrite outgrowth, and axon elongation as well as synaptogenesis [677]. Thereafter, the GT1b, GD1b, Gd1a, and GM1, which are recognized as main brain-specific gangliosides, increase during the most active myelination period ranging from 5 months of gestation to age 5 years. SM and GalCer are also enriched in myelin during the myelin development, and sulfatide and GalCer occupy approximately 4% and 23% of the total myelin glycolipids, respectively [678]. After age 5 years, the GM1 and GD1a levels are decreased, whereas the GM3, GD1b, GD3, and GT1b levels are increased.

GD3 is enriched in the nuclear embryonic brain cells and decreased in postnatal-brain nuclei. GM1 enhances neurite outgrowth in the nuclear membrane [679]. If the simplest ganglioside, GM3 synthase gene ST3GAL5, is mutated, early epilepsy syndrome onset is observed with degenerative motor and cognition [680]. Similarly, the GM2/GD2 synthase gene B4GALNT1 is mutated; hereditary spastic paraplegia syndrome is appeared with neuropathy and dysfunctional intelligence [681]. GM1 is characteristically deficient in Parkinson's disease (PD) and Huntington's disease (HD) [682]. GM1 as a predominant brain ganglioside protectively influences CNS injury. GM1 has neurotrophic, neuroprotection, and anti-neurotoxic activities. Gangliosides neuro-tropically act for neuroprotective effects on neurotrophin signaling, which contributes to the survival of neuronal subpopulations. GM1 potentiates the tropomyosin-related kinase (Trk) receptor activity for neurotrophin release [682]. The mammalian neurotrophins include five families including BDNF, NGF, and neurotrophin-3/4/5. GM1 and GT1b activate neuronal differentiation and dendrite outgrowth by TrkA-enhanced NGF dimerization and NGF phosphorylation potentiates neuronal progenitor cell entrance into postmitotic neuritogenesis and neurons maturation [683]. GD3 also interacts with histone H1 in the nucleus, potentially regulating epigenetic regulation of histone acetylation [666, 684]. In mouse neural stem cells, GD3 binds to EGFR for the self-renewal of neural SC. GD3 contributes to EGFR microdomain localization in membranes. GD3 ablation and depletion reduce the EGFR level and stimulate EGF-elicited EGFR degradation toward self-renewal regression [685]. If GD3-specific autoantibody is generated, it inhibits the recognition or alternatively induces apoptotic cell death to lead to a defected neural function. GD3 ganglioside is present in both stem cells of mouse and neural cells of humans [674].

GM3 and GD3 induce β1-integrin expression for proliferation and renewal of SCs [686]. Glycolipids may regulate gene expression in a manner of cell-type specificity and of stage specificity. Differentiated cells and SCs contain each specialized nuclear envelope protein and specific epigenetic modifications for nuclear spatial localization of genes. It has been hypothesized that nuclear GM1 regulates gene expression in differentiated cell in neuronal region [687]. GM1 in nuclear region recognizes histone acetylation modified on the 5′-flanking region of the GalNAc-T gene promoter of differentiated cells of neurons. GM1 binds to active chromatin region through acetylated histone regions. GM1 also induces the GM2/GD2 synthase gene promoter activity to provide GM2 production [671]. Thus, GM1 allows a positive regulation for neuronal maturation. Enhanced gangliosides in brain modulate neuronal plasticity. Neural development is controlled by differentiation events of GSLs metabolism reprogramming with a switch from globo- to ganglio-series synthesis. Defect in the GSL switch displays malfunctional neurodevelopment in humans, indicating GSLs' role [688]. GSLs reprogramming and neural expression during neural differentiation lead to normal neurodevelopment, but its defects induce neuropathy. The GSL synthesis in brain cells can triggers cell differentiation in a spatial order [689]. In addition, GSLs allow the cells capable to stably maintain the identity, an important property.

Gangliosides are crucial for development and protection of the CNS. Therefore, from the functional roles of gangliosides in neuronal regeneration and NSCs, gangliosides are considered as potential replacement or complementation agents. Gangliosides are extracted from calf brains for neuronal disorders such as CNS lesions and Parkinson's disease. The commercial gangliosides named Cronassial® in Germany and Nevrotal® in Spain are known. Also, the trade named Sygen® in Italy is another example. However, the commercial agents were withdrawn from the markets due to autoimmune response to gangliosides. Several effects of gangliosides are known cancer, diabetes, and infection.

6.2 GD3 and IL-15 Interaction Inhibits T Cell Proliferation and NO Production

IL-15 activation is necessary for the reactivity and physiology of glial cells. Modulation of IL-15 activity has therefore been suggested to create strategies to regulate gliotic pathways upon inflammation. GD3 upregulates the proinflammatory response in IL-15-mediated microglial cells, where microglial cells are regulators in the inflammation in the CNS. GD3 binding to cytokine IL-15 blocks the T cell growth elicited by IL-15 and the NO synthesis and NF-kB signaling [690]. Binding of GD3-specific autoantibodies to GD3 prevents the GD3-IL-15 binding to stimulate the inflammation in the CNS, as indicated in ZIKV infection. The ganglioside neurostatin or IL-15-blocking antibody blockade of IL-15 inhibited the NF-kB signaling pathway and consequently leads to the decreased iNOS expression and NO production. Inhibition of IL-15 signaling also inhibits the MAPKs including ERK1/2 and p38 [691].

6.3 Human Embryo Stem Cell Development

6.3.1 Surface Carbohydrate Markers of Human Embryo Stem Cell Phenotype and Stemness

The stem cell technology is currently applied to explore the therapeutic alternations to cure human diseases but unsolved issues remain. The embryonic SCs (ESCs) of mouse and human ESCs (hESC) [690–692] have been established from the blastocyst mass of inner cells toward regenerative medicine [692]. Historically, Shinya Yamanaka generated the induced pluripotent stem cells (iPSCs) technology through generation of iPSCs from mice and human [693, 694] fibroblasts by retroviral transfection. hESCs and human iPSCs generate cells of each embryonic germ layer of endoderm, mesoderm, and ectoderm, and eventually, differentiate into human cells. Initially, human PSCs were cultured on embryonic feeder cells of mice, raising issues for the immunogenic grafts and rejection antigens, although currently having feeder- and xeno-free solutions. Pluripotent hESCs and human iPSCs will expand to biotechnological applications for therapies. Embryonic stem cell markers of hESCs [692] are carbohydrates including the tumor recognition antigen (TRA)-1-60/TRA-1-81 and SSEA-3/−4. Interestingly, new carbohydrate markers such as the sialyl-lactotetraose antigen and human ABO blood-H type 1 antigenic epitope, which has later been named SSEA-5, are added for hPSC markers [695, 696]. The sialyl-lactotetra antigens and ABO blood group H type 1 antigens are also located on glycoproteins and also on GSLs. However, the SSEA-3 and -4 carbohydrates as forms of globo-series are exclusively attached to GSLs. The SSEA-3 and SSEA-4 are the absolute antigenic biomarkers to detect undifferentiated hPSC, while the SSEA-3/4 GSLs are also found in certain tissues of adult humans [692].

Carbohydrate antigens expressed on cell surfaces are involved in properties of stem cells like differentiation and immunity. The cell surface fate of carbohydrate epitopes easily enable to use in human PSCs. Although the most frequently used biomarkers for the identification of human PSCs are known for globo-series GSLs of SSEA-3 and SSEA-4, currently, there are also limits on our knowledge on human PSCs GSLs due to limited intact cells amounts. The GSLs in human ESCs are relatively well studied in view of the distribution and change of the GSLs during hESC differentiation. The hESCs can be prepared from the blastocyst inner cell mass showing their pluripotency [697], while iPSCs of humans are produced from adult tissue-derived cells [698]. Mammalian cells express a series of surface carbohydrates to function as cell modulators. Certain surfaced carbohydrates on SC are frequently used as definition and characterization molecular markers of ESC. In addition, the ESC carbohydrate epitopes are also used to monitor the level of their differentiation and renewal during culture [699, 700]. Several glycans are found in the surfaces of cell itself and glycoproteins, glycolipids, and proteoglycans. The glycans directly mediate biological events with the regulation of signaling pathways in cells. Not frequently, certain heparan sulfate (HS) mediates cellular functions such as self-renewal, stemness, and differentiation of ESC by signaling molecules such as bone morphogenic protein (BMP), Wnt, and FGF [701]. Certain keratan sulfate (KS)-related structures like podocalyxin KS-related structure of glomerular podocytes act as the hESCs and human iPSCs markers. Podocalyxin is a major pluripotent marker discovered from transmembrane sialylglycoconjugates. Antibodies against TRA-1-60/1-81 bind to the podocalyxin KS-related structures. Cell surface markers are glycoconjugates named SSEAs. The hESCs and iPSCs are distinguished by such glycomic antigens which are highly cell type specific to reflect cellular changes [702].

Human ESC can be transformed even into any derivatives of cell types of three basic germ layers of developing embryos. In addition, human ESC can self-replicate indefinitely. These characteristic cells are the desired human ESC applicable for regenerative medicine but requiring the unique condition that the grafted cells should be tolerant against the recipient immune system. Human PSC includes ESC and iPSCs, which are indefinitely induced for differentiation to all human cell types [703]. The human PSC is immunologically allograft to allogenic subjects, from the expressed histocompatibility antigen, HLA, and the ABO blood type. Human PSC and human ESC and their differentiated cells express A and B antigens in the ABO blood type system [704]. Certain stage-specific embryonic antigens (SSEA) of glycan structures are expressed in mice during embryo development at the early stage [705]. Human ESC lines also expressed blood group type A antigen and type B antigens. For example, human ESC lines of SA121 and SA181 express the ABO blood group type A antigen and ABO type H antigen on type 1 core lactotetraose of GSLs, but not the Le^aLe^b antigens [706]. AB(O)H antigens of adult human erythrocytes are attached to glycoproteins [707]. However, the AB(O)H attached to GSLs are linked to type 2 core chains [708].

When sialyl-lactotetraose (s-Lc4) and TRA-1–60 are compared to the SSEA-3/SSEA-4 structures, the SSEA-3/SSEA-4 structures as the ESC marker are rather suboptimal. SSEA-3 is broadly and weakly stained in human PSC [697], while SSEA-4 is abundant in the human PSC [709]. The expressions of the blood group type 1 H structure and SSEA-5 structure are gradually reduced and consequently disappeared in human PSC during differentiation. Therefore, the ABO blood H type 1 epitope and SSEA-5 antigen epitope are defined as a carbohydrate biomarker of undifferentiated cells [710]. The s-Lc4 is now used as a pluripotency marker during differentiation process [696, 697]. One more specific embryonic antigen, SSEA-1 or CD15 and Le^x, is specifically abundant in mouse, but not in human PSC [689]. The wrong conception is caused by the origin that the SSEA-1-recognizing antibody (clone MC48032) is raised against the Le^x epitope linked to extended type 2 core chains [711]. This specific property implies for the Le^x antigen expression, not SSEA-1, on human PSC. Pluripotent stage cells do not express SSEA-1. The exact biological definition of human ESC and human iPSC is still debated [712]. In addition, the compositions of GSLs are largely similar in the two human PSC types of human ESC and human iPSC. However, GD1a and GD1b levels are reduced and GM3 levels are increased in human iPSC, but not human ESC [697].

In the clinical application of human PSCs of human ESCs and human iPSCs, teratoma formation issue is the current threat, which is raised by residually resided undifferentiated cells. To solve this concern, a recent Mab has been developed. Using the Mab reactive to human ESCs, a new type SSEA-5 to recognize human PSCs' H type-1 glycans, the teratoma-forming cells from heterogeneously differentiated cells were eliminated. To remove teratoma-forming cells, human PSCs-specific pluripotency surface markers have been discovered during differentiation. The identified antigens include CD9, CD30, CD50, CD90, and CD200. Immunological depletion of the SSEA-5 and PSMs using Mabs specific for SSEA-5 and PSMs perfectively eliminated teratoma-forming cells from incompletely differentiated human ESC cultures or undifferentiated human ESCs [710] (Fig. 6.1).

As a mouse model of human ESC, alternatively usable mouse ESC has been established; however, they are at developmentally different stages, compared to human ESC. This different developmental stage found between human ESC and mouse ESC indicates their respectively different signaling mechanisms. In the mouse ESC, for example, their self-renewal or stemness event is involved in signal transducer activator of transcription (STAT3) and leukemia inhibitory factor (LIF) axis responses, but not for human ESC self-renewal [713]. BMP4 and Smad signaling with LIF are also involved in the mouse ESC self-renewal and stemness [714]. However, BMP signaling induces differentiation of hESCs [715], while Wnt/b signaling induces self-renewal in both mouse ESC and human ESC [716]. However, activin, fibroblast growth factor 2 (FGF2), and nodal axis involve uniquely in the stemness and maintenance of self-renewal of human ESCs [717]. Therefore, mouse ESC and human ESC keep and maintain their pluripotencies using each specific signaling factor. Human ESC under the primed state can be converted to a naıve state, which is similar to the mouse ESC. Cell surface glycans regulate the distinct signaling for self-renewal maintenance in mouse

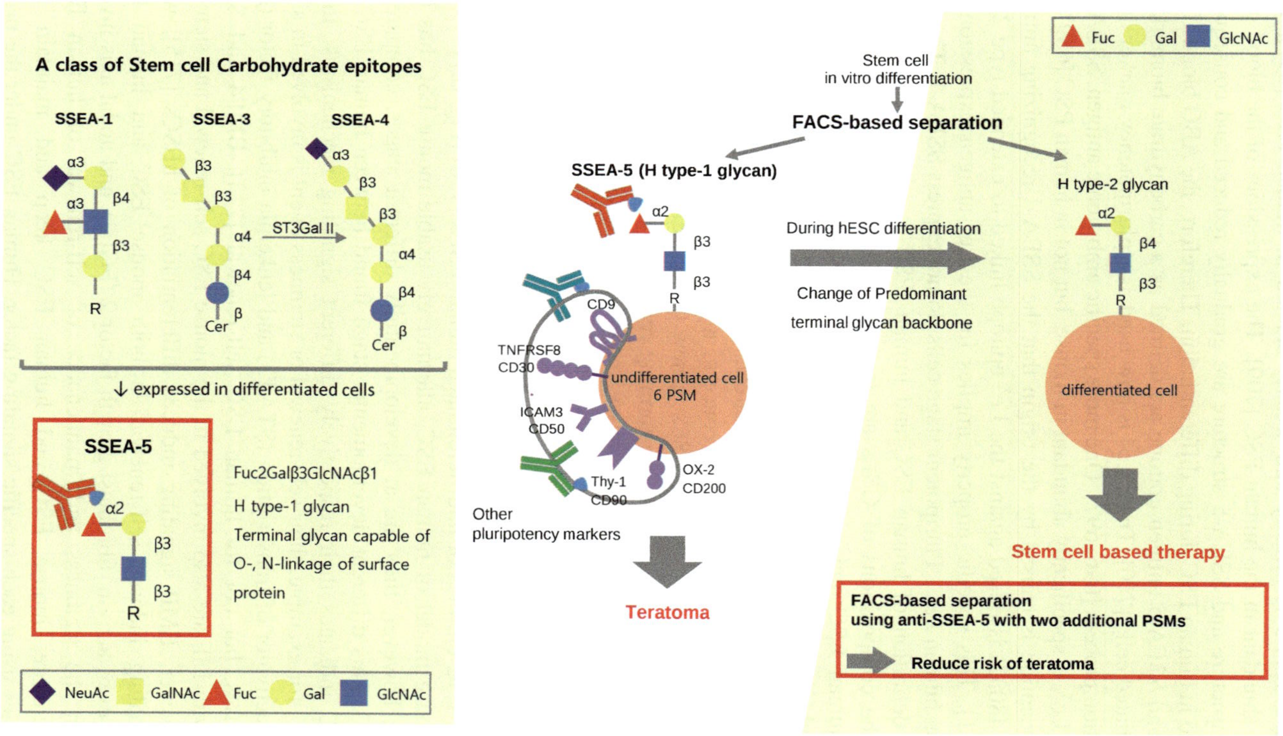

Fig. 6.1 Human pluripotent stem cell (hPSC) replacement by SSEA-5-specific antibody removes teratoma-forming cells during differentiation. (Adapted from Ref. [710] Tang et al. 2011. Nature biotechnology 29 (9), 829)

ESC and human ESC. For example, differentially expressed cell surface glycans such as LacdiNAc (GalNAcβ1,4GlcNAc) motif on gp130 and LIFR in primed state cells and naıve state cells are differentially responded to STAT3 and LIF signaling of mouse ESC (naıve state) and human ESC.

LacdiNAc motif linked to glycoproteins and glycolipids as shown in glycoprotein hormones of vertebrates is synthesized by catalysis of β4GalNAc-T3 enzyme [718–720]. The biological roles of LacdiNAc are not clarified yet. The mouse ESC and human ESC show different responses to exogenous signals due to the cell surface glycans of LacdiNAc. The mESCs shRNA-knocked down cells in the β4GalNAc-T3 gene, which encodes the LacdiNAc synthesizing enzyme has been checked for the selfrenewal capacity [721]. Silencing of β4GalNAc-T3 gene decreases in the level of LacdiNAc synthesis. The ESC marker alkaline phosphatase positiveness is decreased. ESC-maintaining Oct3/4, Sox2, and Nanog levels are also reduced. The growth level is also reduced compared to the normal mESC. Thus, LacdiNAc is an essential self-renewal factor of mESCs. Next, to see how LacdiNAc regulates signaling, LIF, BMP4, and FGF4 were treated with the cells. Among STAT3, Smad1, and ERK, STAT3 phosphorylation only was decreased, indicating that LacdiNAc regulates LIF/STAT3 signaling. LIFR and gp130 expression were not changed. However, LacdiNAc interferes with LIFR-gp130 heterodimerization. The LIFR and gp130 are located on the PM lipid raft of cell surfaces. LacdiNAc strictly leads to the accurate cellular distribution and localization of LIFR and gp130 on cells. LIFR and gp130 contain LacdiNAc residue, as confirmed with its lectin WFA. In addition, LIFR and gp130 associated in caveolin-1 bind to the lipid raft microdomain via LacdiNAc. The lower response of mEpiSC or hESC to LIF is due to LIFR and gp130 LacdiNAc levels, because these cells lowly synthesize LacdiNac, caused by low β4GalNAc-T3 expression than mESC. LIF induces a return to naïve state; mEpiSC generated from β4GalNAc-T3-KD cell returned to naïve state. Thus, LacdiNAc is necessary to induce and maintain mESC. LacdiNAc specifically expressed in mESC is expressed in LIFR and gp130 in the lipid raft. Thus, self-renewal of mESC can be induced by LIF/STAT3 signaling.

6.3.2 GSLs of Human ESC

Glycans and GSLs localized in the cell surfaces are extracellularly exposed. The GSLs expression is controlled quantitatively and qualitatively by each species. The ceramide region contains a fatty acid-amide linked base. Ceramide and carbohydrate moiety are combined to a numerous structural diversity for more than 400 compounds [722]. GSLs are classified into two groups of negative-charging acidic GSLs and neutral non-acidic GSLs. The acidic GSLs are subclassified into the SA-containing gangliosides and sulfate-esterified ceramides like sulfatides. Also, GSLs are subclassified, depending on the core carbohydrate structures. In humans, the most common non-acidic GSLs include the lacto-type 1 structure with the Galβ1,3GlcNAc sequence, neolacto-type 2 structure with Galβ1,4GlcNAc sequence, and globo-type 4 with Galα1,4Gal core structures. The abundantly occurring gangliosides include the structures of ganglio with Galβ1,3GalNAc structure or neolacto core structure. Glycoproteins also consist of the lacto and neolacto core structures. Ganglio and globo core structures are uniquely present in GSLs. GSLs are fractionated into non-acidic forms, gangliosides and sulfolipids. Monoclonal antibodies can recognize SSEA-3 and Globo H.

GSLs are essential for embryo development in mammals. Cells are decorated with a glycocalyx. The specialized structure glycocalyx contains GSLs, glycoproteins, and proteoglycans. Changes in GSLs expression profile designate phenotype changes during mammalian embryogenesis. Carbohydrate antigens in ESC have mainly been identified so far by experimental reagents of monoclonal antibodies or lectins. These reagents cannot discriminate between protein-linked glycans and lipid-attached glycans. Certain structured GSL-specific antibodies designated as SSEAs are experimentally utilized in differentiation phenotype characterization of ESCs. Changes in surface GSLs are frequently detected in the ESC differentiation. GSL profiles in human ES cells are well clarified. ES markers such as lacto series and globo series GSLs such as Globo H, Gb4Cer, fucosyl Lc4Cer, Lc4Cer, and disialyl Gb5Cer as well as SSEA-3 and SSEA-4 are found in the undifferentiated ESCs of human and iPSCs of humans. In the differentiation to embryoid phenotypes, the GSLs are shifted from lacto-series and globo-series GSLs to ganglio-series. In the ESCs to neural progenitor differentiation, GSLs structures shift to ganglio-series of disialylganglioside GD3. During endodermal differentiation, GSLs structures shift mainly to Gb4Cer and minorly to SSEA-3 and SSEA-4 or GD3 [723].

SSEA structurally belongs to globoseries GSL. For the known hESC GSLs, non-acidic GSLs of the lacto series including ABO group H type 1 pentaosylceramide and type 1 core chain of lactotetraosylceramide and the globo series of globo H hexaosylceramide, globotetraosylceramide, and globopentaosylceramide/SSEA-3 are included. Additionally, GM1, GM3, GD1a, or GD1b gangliosides, disialyl globopentaosylceramide, and sialyl-globopentaosylceramide known as SSEA-4 are also included. For more precise details, 9 years ago in 2011, it has been found the 4 carbohydrate markers of ABO blood H determinant with the type 1 core structure, globo H, globopentaosylceramide known as SSEA3, and sialyl-globopentaosylceramide known as SSEA4 from the GSLs composition of two human ESC lines of H9 and HES5 [94]. Moreover, from per-methylated lipids, the non-acidic GSLs of globotetraosyl-Cer, lactotetraosyl-Cer, H type 1 pentaosyl-Cer, globopentaosyl-Cer/SSEA3, Globo H hexaosyl-Cer, GM3, sialyl-globopentaosyl-Cer/SSEA4, disialyl-globopentaosyl-Cer, GD1a, and GD1b were further found. Separately to the above, several type 2 core GSLs including *neo*-lactotetraosyl-Cer, H type 2 pentaosyl-Cer, Le^x pentaosyl-Cer, and Le^y hexaosyl-Cer were found with the blood group A type 1 hexaosyl-Cer. Moreover, other mono-, di-, and triglycosylceramides were also found with GalCer, GlcCer, LacCer, galabiaosyl-Cer, globotriaosyl-Cer, and lactotriaosyl-Cer [724]. Our current information of cell surface carbohydrates on ESC has come from studies of mouse embryonic cells. However, from the undifferentiated hESC, GSLs of the globo-series including globotetraosyl-Cer, globopentaosyl-Cer/SSEA-3, and Globo H hexaosyl-Cer as well as the type 1 core structures of Lac-tetraosyl-Cer and Fuc-Lac-tetraosyl-Cer/H type 1 pentaosyl-Cer were also found. In addition, from the undifferentiated human ESCs, GD1b, GD1a, GM1, or GM3, disialylglobopentaosyl-Cer and sialyl-globopentaosyl-Cer known as SSEA-4 were also found. Interestingly, the differentiated neural progenitor cells from hESC showed the ganglio-series GSLs [94], whereas the differentiated endodermal cells display predominantly globotetraosyl-Cer [725].

The synthesis of GSLs depends on the GSL synthesizing enzymes, as they are tightly regulated during development. The glycosylation patterns of hESCs and hiPSCs are clues of cell differentiation and stemness. GSL changes during differentiation of hESs and iPSCs into various derivatives. Embryo cell surface molecules in a stage-specific manner regulates cell-cell interaction and differentiation during early development. Such changed antigens-recognizing mAbs display stage specificities to recognize carbohydrate epitopes. In the undifferentiated human ESC, cell surface marker profiling covers SSEA-3/4 and the keratan sulfate (KS)-associated antigens such as TRA-1-60 or TRA-1-81 [701, 726–728]. SSEA-3 known as Gb5Cer with the Hex-HexNAc-Hex-Hex-Hex-Cer glycan structure and SSEA-4 known as sialyl-Gb5Cer (NeuAc-Hex-HexNAc-Hex-Hex-Hex-Cer) glycan structures are all GSLs (globopentaosyl-Cer and sialyl-globopentaosyl-Cer, respectively), since the globo-series glycans are only found in GSLs.

SSEA-4 is the prominently major biomarker of human PSC. Antibody binding to SSEA-4 would be the most basic approach to immunologically sort-out undifferentiated cells [729] for therapeutic application. Among SSEAs, SSEA-1 reacts with neolacto-series such as GSL Le^x, whereas the makers of SSEA-3/−4 recognize Gb5 form and monosialyl-Gb5 form, respectively. The known human ESC-specific biomarkers include SSEA-3, SSEA-4, and certain GSL forms such as globo-series and lacto-series of Gb4-Cer, Fuc-Lc4-Cer, Lc3-Cer, Globo H known as fucosyl-Gb5-Cer (Fuc-Hex-HexNAc-Hex-Hex-Hex-Cer) structure and disialyl-Gb5-Cer [730]. Switching shift of core carbohydrate structures from the globo-series GSL types and lacto-series GSL types to the ganglioside types frequently occur during ESCs differentiation. GSL expression pattern during early development of embryonic cells shifts from globo-series types to neolacto and lacto-series types, and thereafter to ganglio-series GSLs. Such switching shift is collaborated with the changed expression of glycosyltransferases during differentiation. Glycosyltransferases of GSL synthesis are important and the enzyme gene knockout animal models display specific developmental defects [522]. Multivalent Le^x-Le^x epitope interaction, termed CCI, leads to cell aggregation. Same shifting from the globo-series GSL types and neolacto/lacto-series GSL types to ganglioside GSL types is also found in the differentiation process of embryonal carcinoma cells (ECC) and ESCs of humans by surface glycoconjugates [731].

When compared to somatic cells of humans, the stem cells of humans exhibit the higher levels of the GSLs of globopentaosyl-Cer/SSEA-3, sialyl-globopentaosyl-Cer/SSEA-4, Fuc-Lac-traosyl-Cer/SSEA-5, difucosyl-neolactotetraosyl-Cer, neolactopentaosyl-Cer, neolactotetraosyl-Cer, globotetraosyl-Cer, the NeuGc-GM1 ganglioside, and gangliotriaosyl-Cer. Therefore, stem cell GSLs are contaminated with the non-human SA for NeuGc-GM1. The hESC GSLs initially include Globo core chains (globotetraosyl-Cer, globopentaosyl-Cer/SSEA-3, Globo H hexaosyl-Cer, sialyl-globopentaosyl-Cer/SSEA-4, or disialyl-globopentaosyl-Cer) and type 1 core chains (lactotetraosyl-Cer or fucosyl-lactotetraosyl-Cer/ABO blood group H type 1 pentaosyl-Cer) with the GM1, GM2, GD1b, or GD1a gangliosides. A similar composition was found in hiPSC GSLs [697]. Minor GSL forms like type 2 core chain GSLs and unusual forms like acidic GSLs of sialyl-globotetraosyl-Cer and sulf-globopentaosyl-Cer are also known.

Regarding the relationship between carbohydrate expression of undifferentiated and differentiated cells, SSEA-3/4 maintains hESC pluripotency. From the hESC depletion of GSLs differentiates ectodermal, endodermal, and mesodermal cells; it is considered that GSLs are not critical for maintaining the undifferentiated hESC status. However, in the embryoid bodies generated from hESC, GSLs function cellular differentiation. GSLs in embryonic development and tissue differentiation regulate cell-cell and cell-ECM recognitions. Because hPSC cell replacement therapy has the tumorigenic risks from residual undifferentiated cells, which hPSC cells are still remained in the differentiated cells [732]. Therefore, the residual undifferentiated hPSC contamination should be checked and removed in the differentiated cell population. hPSCs surface glycan biomarkers eliminate undifferentiated hPSCs [729]. The Mab specific for the H type 1 GSL of hiPSC will be the case of selective removal of undifferentiated cells. The biomarkers composed of sialyl-lactotetra structure in the undifferentiated hPSC [697] are also meaningful for removing undifferentiated stem cells.

Interestingly, several GSL types in human ESC are largely expressed in many human tumors [733], indicating that such carbohydrate profiles are correlated to both cancer cells and ESC toward eternal cell proliferation [734]. These glycans are biomarkers for hESC and cancers [94]. However, some monoclonal antibodies specific for the GSLs are reactive against carbohydrate glycan epitopes [735]. The current interest is in the antibody specific for SSEA-3/globopentaose, named antibody MC631 specifically used for human ESC [701], because this antibody binds to globotetraosyl-Cer, globo H hexaglycosyl-Cer known as fucosyl-Gb5Cer with the Fuc-Hex-HexNAc-Hex-Hex-Hex-Cer structure, and H type 1 pentaglycosyl-Cer known as fucosyl-Lc4Cer with the Fuc-Hex-HexNAc-Hex-Hex-Cer structure. Therefore, specific antibody development against each carbohydrate epitope is timely required for the precise carbohydrate antigen determination as disease biomarkers or hESC biomarkers. Aberrant glycosylation is a cancer phenotype and thus tumor-associated surface carbohydrates can be applied for targeting in cancer immunotherapy. For example, GD2-targeting immunotherapy is applied for neuroblastoma as the first trial for antiglycan antibody. Globo H-targeted immunotherapy of breast cancer is also such case, indicating future use of glycans for tumor immunotherapeutics [730] (Fig. 6.2).

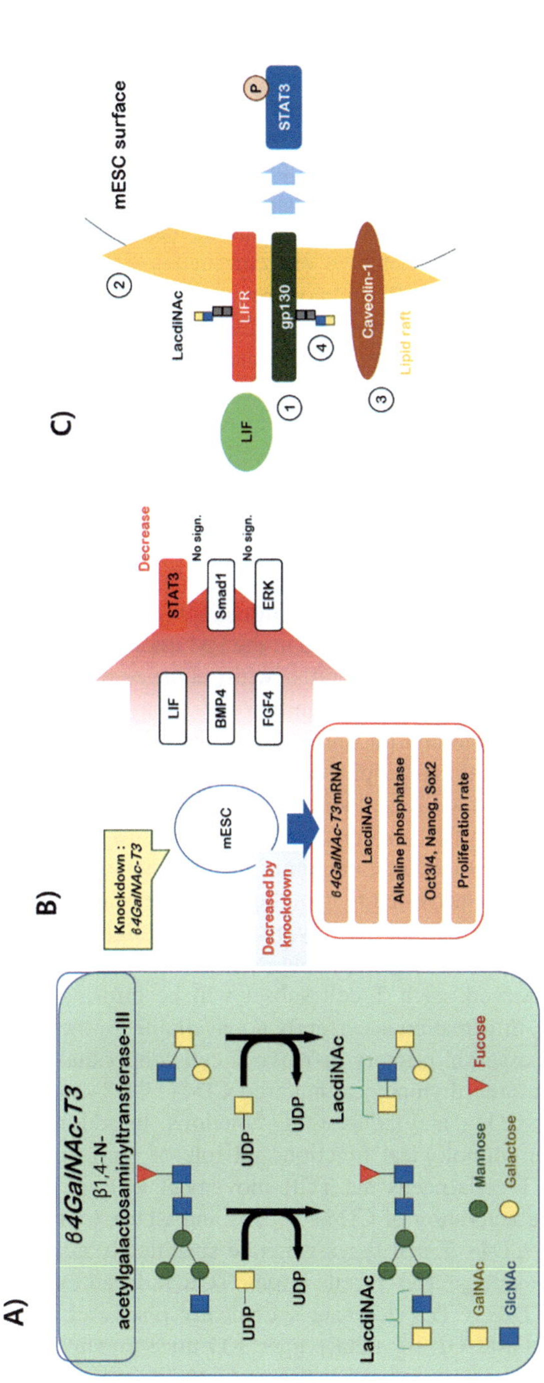

Fig. 6.2 LacdiNAc structure of GalNAcb1-4GlcNAc maintains self-renewal of mESC, stemness by leukemia inhibitory factor ILIF/STAT3 signaling. (Adapted from Ref. [721] Sasaki et al. 2011. Stem Cells. 29(4), 641-50)

6.4 Ganglioside and CD4+ T/CD8+ T Cell Activation

Activation of T cells is a prerequisite to respond to antigens in the immune system. For the activation of T cells, prerequisite action needs the T cell receptor (TCR) movement to PM lipid rafts, where gangliosides associated with lipid rafts. Lipid rafts engage in the TCR-elicited T cell signaling. The T cell PM forms the versatile cell surface receptors affordable for the functional flexibility possible for T cell activation. There is a big mismatch in our current knowledge regarding the PM function and components [614]. The major components of cholesterol and GSLs of the PM lipid rafts are required for immune cell functional activation [736]. In T cells, cellular PM cholesterol and GSLs form TCR clusters and regulate TCR activation or TCR function [737–739] including cell death, apoptosis, and recycling endocytosis of cellular receptors. Distinct PM GSL profiles form or deform Th cell subpopulations of Th1, Th2, and Th17 that define T cell phenotypes. T cells having high cellular membrane order stably constitute immunological synapses to attract a Th 2 phenotype cells while T cells having low cellular membrane order form unstable immunological synapses with proinflammatory phenotype [740, 741].

The term microdomain or "lipid rafts" was at first established by a study of Kai Simons and Elina Ikonen, from the conceptional sphingolipids and cholesterol as complexed signaling molecules embedded in the PM microdomains [742]. It is well known that GSLs-associated lipid rafts with cholesterol and sphingomyelin (SM) activate T cells through recognition of adaptor signaling molecules [737]. Changes in PM lipid and phospholipids composition as well as distribution influence the physiological and biophysical properties of lipid rafts, defining T cell phenotypes. As evidenced, GSLs influence T cell subset differentiation and function as phenotypes such as the TCR-mediated signaling [738, 739, 743, 744]. The T cell PM surface receptors for functional outcomes include induction, growth, and cytokine synthesis. T cells are activated by gangliosides. The lipid rafts are functionally essential for T cell differentiation, motility, maturation, migration, and activation in subsets of $CD4^+$ T/$CD8^+$ CTLs. If changes in composition and distribution of gangliosides are observed, each T cell subset will be informative for functional definition. However, information is currently not available for the detailed structures of gangliosides expressed in primary or naïve T cell subpopulations. Gangliosides expressed in non-matured thymocytes, matured CD4+, CD8+, and primary T cells define characteristic CD4+ and CD8+ CTL functions. In addition, the expression profiles define the immunological function and role of each type of T cells. The activation event of T cells needs the TCR movement to lipid rafts composed of gangliosides. During activation of CD4+ T cells and CD8+ CTLs, specific gangliosides are required. CD4+ T cell types generate specifically a-series gangliosides. For example, GM3 synthase KO T cells impair TCR-induced cytokine release and clonal expansion. CD8+ CTLs generate specifically o-series gangliosides. CD8+ CTLs derived from GM2-GD2 synthase gene KO mice impair TCR-elicited cytokine release and clonal expansion. TCR-driven CD4+ T cells activation also needs ganglioside a-series. Also, TCR-driven CD4+ and CD8+ CTLs activation needs distinctly a-series and o-series, respectively.

Profiles of the ganglioside synthesis during T cell differentiation and activation should be distinct for each ganglioside species [743]. The T cell activation is initiated as a first step by the TCR movement to PM lipid rafts. Gangliosides in lipid rafts molecules are important for T cell functional activation and thus different T cells types require and display distinct gangliosides for the activation. Such T cell subpopulation is applicable for treatment of immune diseases [744]. For example, upon activation of $CD4^+$ T cells as central adaptive immune cells, they are ready to proliferate, migrate to inflamed tissue sites, and acquire Th cell functions. Then, the Th cells consequently function for the downstream immune responses such as inflammation and cytokine production against infection manifestation and malignant growth [745]. PM lipid profiles of GSLs and cholesterol concentration define Th cell subsets of Th1, Th2, and Th17 [738–740, 746–748] as functional T cell phenotypes.

T cells lead to immune surveillance of hosts and antigen recognition. Matured T cells normally produce either $CD4^+CD8^-$ phenotype-specific antigens or $CD4^-CD8^+$ phenotype markers on surfaces. $CD4^+$-helper T cells functions include humoral immunity or allergic responses. In contrast, $CD8^+$ cytotoxic T cells exert tumor cells- and virus-infected cells-killing activity. In T lymphocytes, PM lipid rafts control TCR signaling with the well-known coreceptor of CD4 and CD8, as well as signaling adaptor proteins including Src kinase family of Fyn and Lck, and transmembrane adaptor protein linkers as well as protein kinase C (PKC) [749]. PM gangliosides also involve in the activation of T cells. In fact, the TCR is normally clustered with CD3 and CD28 and then further polarizes GM1a in $CD4^+$ T cells only. However, TCR does not polarize GM1a in $CD8^+$ T cells [745]. During T cell activation, GM1a-associated TCR clustering with CD3 and CD28 is found in polarized CD4+ T cells; however such clustering is not found in the polarized CD8+ CTLs. During T cell polarization of humans, GM1a and GM3 are embedded in lipid rafts microdomain to reorganize the functional microdomain [750]. In the T cell activation, the TCR-CD3 microdomains play key roles such as membrane trafficking, cytoskeletal organization, cell adhesion, pathogen attachment, entry, and signaling [751]. The TCR-CD3 complexed microdomain also elicits the immunologic synapse function, recruiting signaling molecules to TCR-CD3 microdomains and activating T cells. Lipid rafts form the immune synapses in the interphase of APCs and T cells. GM1 and GSL are the major lipid rafts components to form GM1 clustering with TCR-CD3. PKCθ functions in TCR-CD3 activation signaling. TCR-CD3 interacts with GM1 before T cells activation on cell surface. T cell activation occurs in the TCR-CD3 clusters interfaced with APCs. During activation of T cells, Ag-specific TCR-CD3 is clustered with CD4- or CD8-expressed co-receptors [752]. GM1-embedded microdomains elicit the signaling for TCR-CD3-derived activation after GM1 clustering, colocalization, and interaction with TCR-CD3. TCR-CD3 and GM1 are associated during T cells activation. CD3- and CD28-reactive Ab treatment elicits the formation of TCR-CD3 and GM1 nanodomains.

The gangliosides remodeling is functionally regulated for other cell types. Considering the impact of GM1 on $[Ca^{2+}]i$ in human T lymphocyte (Jurkat) cells [753], the GM1 presentation by effector T cells and Galectin-1 expression in regulatory T cells upon activation suppresses autoimmune disease state [754]. In general, $CD4^+$ T cell subsets produce predominantly a-series gangliosides among GSLs, while $CD8^+$ CTLs produce predominantly o-series ganglioside GSLs. Seemingly, ganglioside a-series mainly activate the TCR-dependent $CD4^+$ T cells. During the $CD8^+$ CTL activation, o-series gangliosides are required. Distinctly different gangliosides produced by $CD4^+$ T cells and $CD8^+$ CTLs may determine directions of specific T cell subset immunity. It was proposed that cellular differentiation repertoire shift from premature or immature type T cell types to matured type of T cells is modulated by defined GSL production of the distinct T cell subsets.

The GM1 polarization is observed uniquely in $CD4^+$ T cell type but not in $CD8^+$ CTL type. During the T cell subpopulation onset, TCR clustering is stimulated by treatments with antibodies specific for CD3 and CD28. Different ganglioside profiles in CD4+ and CD8+ CTL T cells decide their distinct immune functions. CD4 and CD8 interact with ganglioside-embedded lipid rafts. A co-stimulatory CD28 distinguishes the two cell types of CD4+ T cells and CD8+ CTLs. For the CD4+ T cell types, the CD28 induces the formation of the immune-synapses which CTx-B is associated with in lipid rafts through PKCθ [755]. GA1-positive CD8+ CTLs activate the TCR in CD40/CD28-mediated allograft rejection [756]. Thus, each CD4+ and CD8+ CTLs bear a specific lipid rafts-embedded ganglioside profile for Tc and Th effector cell functions. GA1-expressing CD8+ CTLs generate high IFN-γ level during TCR activation rather than GA1-deficient CD8+ CTLs. Clonal expansion of CD8+ CTL cells and allograft rejection are inhibited by GA1-mAb injection [755]. In mouse CD4+ T cells, T cell activation is induced by a-series gangliosides. Human CD4+ T cells mainly produce GM3 for TCR signaling. Mouse CD4+ and CD8+ T cells upon CD3 MAb treatment differently synthesize N-glycans of glycoproteins [757]. The differently synthesized N-glycans present on TCR are found in CD4+ T cells and CD8+ CTLs, influencing TCR-activated function of T cells [758]. The co-receptor proteins of CD4 and CD8 recognize the complex of TCR-MHC depending on glycosylation. Gangliosides bind to N-glycans and gangliosides on the CD4+ and CD8+ CTL T cells membranes recognize the TCR for TCR signaling. Activated human T cells express GM1a. GM1a content is enhanced in the CD4+ T cells only. GM1a ganglioside is dominantly produced in the self-antigen-reactive CD4+ T cells [759], where Th1, Th2, and Th17 as well as Tregs are included. In the allergic inflammatory process of airway track tissues, GM3 synthase gene-deficient KO mice are not induced for asthma, as the ratio of Th2 effector cells and Treg suppressive cells is shifted to Th2 dominant phenotype [760]. The Th17 cells induce steroid-resistant asthma, neutrophilic and Th2-induced inflammation of airway. GM3S KO mice decrease Th17 cells population number and level. Induction of CD4+ T cells needs ganglioside GM3, while activation of CD8+ CTLs does not need the GM3 ganglioside. For activation of CD4+ T cells, not for CD8+ CTLs, GD3 is needed. Therefore, a distinct type of ganglioside biosynthesis is crucial for each activation of CD4+ T and CD8+ CTL T cells. Moreover, ganglioside biosynthesis defines and determines host immune directions through various T cell subpopulations. The selection event of each specific GSL species should be a directing factor of functional PM lipid rafts in matured type of T cells [743].

6.5 GD1b and Interleukin-2 Interaction

The gangliosides shed on the cancer cell surfaces play important roles in immunosuppression. The acidic GSLs inhibit lymphocyte growth and disturb growth factor-like IL-2 functions by direct IL-2 interaction with gangliosides [761]. The gangliosides are a target in IL-2 binding. For example, using the labeled ^{125}I-IL-2, the binding between IL-2 and gangliosides in aqueous solution has been demonstrated giving a hydrophilic interaction, not hydrophobic interaction. Considering that ganglioside carbohydrates alone cannot bind to IL-2 itself, such aqueous conditioned interaction suggests the possibility of direct binding of gangliosides with IL-2. However, the IL-2 binding to PM bilayer gangliosides depends on the synthesized GSL composition. For more specific recognition of IL2 with gangliosides, a solid matrix immunoassay has been used and the recognition between IL-2 and gangliosides has been demonstrated. Although human IL-2 specifically recognizes GD1b, IL-2 does not bind to other ganglioside types such as Gt1b, GD3, GD2, GD1a, GM3, GM2, and GM10. The specificity of the IL-2-binding capacity of GD1b is interesting in terms of IL-2 ratio. The mechanism how the carbohydrate moiety of the GD1b recognizes IL-2 is interesting to note. The recombinant IL-2 therapy is currently performed for the systematic administration to stage IV melanoma patients. If the patients express GD1b, it is considered that GD1b-expressing tumor or the body fluid GD1b in the tumor patients may prevent the IL-2 therapeutic effects due to the GD1b-IL-2 interaction, raising problems in antitumor-immune response.

Gangliosides are also expressed in non-neoplastic cells, but gangliosides in cancer cells are specifically expressed as the tumor-associated phenotypes in melanoma, metastatic colon, and pancreatic cancer cells [762, 763]. For example, some metastatic colon carcinoma and pancreatic adenocarcinoma cells largely express GM2, GD1b, and GT1b even in the body vascular and lymphatic node fluids of cancer patients [762–766]. Abnormally overexpressed serum gangliosides indicate a poor prognosis [763], because the serum ganglioside levels are increased, as derived from the cancer cells. The poor prognosis of tumor is, therefore, caused by immunosuppression induced by the tumor-associated gangliosides because serum gangliosides are known to function as immunomodulators [767]. In addition, to escape from the immune surveillance, some gangliosides enforce to downregulate CD4 expression of helper T cells [739, 743, 768].

On the other hand, in terms of IL-2-IL-2 receptor (IL-2R) signaling, when IL-2 binds to gangliosides such as GD1b, the GD1b blocks the interaction between IL-2 and IL-2R [769]. For example, it has been known that GD1b inhibits IL-2-elicited growth of the CTLL-2 and HT-2 cells, which were treated with IL2, through the mechanism that inhibits the downstream IL-2 signaling pathway. Gangliosides competely bind to IL-2-binding sites with administered rIL-2 in the clinical application. IL-2 recognizes directly GD1b only, not other simple glycolipids or acidic glycolipids. The disialyl gangliosides such as GD3, but not monosialyl gangliosides such as GM3, stimulate cytochrome C release from mitochondrial membrane during GD1b-IL-2 recognition.

6.6 GM1 and IL-4 Interaction

Ganglioside GM1-binding sites in IL-4 have also been studied. Kholodenko et al. [770] have recently analyzed the GT1b-recognizing sites on the IL-4 using the photoaffine labeling. Determination of the GT1b-binding site in the IL-4 can expand to development of future targeting drugs for interleukin. For photoaffinity binding of labeled GT1b to IL-4, new radioactive and photoaffinity labeled probe on the basis of GM1 modified with the diazocyclopentadien-2-ylcarbonyl (Dcp) group in the head lipid moiety, ^{125}I-Dcp-GM1. Photolysis of Dcp-GM1 forms the ultimately reactive intermediate carbene interactive with adjacent molecules. Because the shedded gangliosides are detected in the bloodstream, they modulate the innate immune cells. The shedded GSLs also inhibit the proliferation of T lymphocytes and some GSLSs inhibit the proliferation of cytokine-dependent T lymphocytes [771], giving an application clue to GSLS-mediated inhibition of normal nonmalignant cells. The cytokine-dependent T cell inhibiting activity of gangliosides increases in the course of GT1b > GD1a > GM1 > GM3. In more specific kinetic mechanism, GM3 and GT1b act as competitive inhibitors of proliferation, whereas GD1a and GM1 simultaneously act as both competitive and noncompetitive inhibitors [772]. The molecular mechanisms how T cell proliferation is inhibited are explained from the fact that the GSLs directly interact with IL-2 and IL-4. Thus, GSLs prevent cytokine-bound T cell activation and provoke immunosuppression. Some gangliosides complex with IL-2 and IL-4 with high-binding affinities [773], although the ganglioside-binding site in the IL-4 is interesting. Recently, by using a photoaffinity labeling strategy, Shapiro et al. [774] developed a photoaffinity analogue of GT1b to find the GT1b-binding site on the tetanus toxin.

To analyze the binding sites of GM1-IL-4 and photoaffinity labeling of ^{125}I-Dcp-labeled GM1, IL-4 is incubated with ^{125}I-Dcp-GM1 and the labeled mixture was subjected to photolysis. The covalently bound IL-4-GM1 derivative is degraded by the endopeptidase Glu-C. The photoaffinity GM1-protein binding forms the cyclopentadiene carbonyl residue (cp)-carrying molecule. It is known that the polar head of GSL is involved in the binding to proteins. Therefore, to understand the binding site of the receptors such as IL-4, the presumable binding site location with the polar head of GSLs is expected as the interaction site. GM1 competitively and noncompetitively inhibits the proliferation of cytokine-dependent T cells [772]. The competition between the ganglioside and cytokine receptor suggests that ganglioside binding and receptor-binding sites on the cytokine are overlapped.

The IL-4 amino acids sequence are the hydrophilic E9 and R88 as the IL-4R-α-recognition site [775]. The hydrophobic moiety of the ganglioside prevents the interaction between IL-4 and the receptor subunit. Gangliosides inhibit the proliferation of the IL-4 responsive CT.4R cells [772]. GT1b is a competitive inhibitor of proliferation, while GM1 and GM3 are moderate.

6.7 GM4 (Neu5Acα2,3Galβ1-Cer) and IL-1β Interaction As Well as Neu5Acα2,3Galβ1,4GlcNAcβ1,2 and IL-1α/IL-4/IL-6/IL-7 Calcium-Independent Interaction

The IL-glycan recognition as the lectin-like specificity is also observed in the condition of 5 mM EDTA, which is the Ca^{2+}-independent carbohydrate-binding mode [776]. Carbohydrate-binding ILs include IL-7, IL-6, IL-4, IL-1α, and IL-1β as a calcium-independent interaction [777]. For example, IL-1α is indeed a calcium-independent carbohydrate-interacting lectin specific for the Neu5Acα2,3Galβ1,4GlcNAcβ1,2 sequence. IL-1β also has a carbohydrate-interacting lectin domain. IL-1β recognizes only the monosialoganglioside of GalCer, GM4 [778]. GM4 is a simple ganglioside having a structure of Neu5Acα2,3Galβ1-Cer, where GM4 is a Gala-series glycolipid. Gala-series glycolipid includes GalCer, sulfatide (SM4s), and GM4. GM4 is mainly expressed in the central nervous systems in myelin of mammalian brain and the erythrocytes [779], but the physiological function of GM4 is still not clear. The IL-1β-recognition is therefore inhibited by similar structured oligosaccharides to Neu5Acα2,3Galβ1-Cer [777]. The long-chain lipid moiety of GM4 seems to be crucial for IL-1β recognition and interaction. IL-1β completely functions compared to the homologous IL-1α, although IL-1α/−1β utilizes the common conserved receptors; however, they exert different biological functions to the receptor signaling during the action. Apart from IL-1α and IL-1β, the other ILs have also binding affinities toward carbohydrate ligands, suggesting that the human ILs may target the independent and different receptors from the typical or classical IL receptors. Such endogenous ligands of the different ILs are not found yet.

For immunological function of carbohydrate-binding ILs, the regulatory effects in the human immune system have been considered, as specific carbohydrate substances can be used as immune therapies. Actually, ILs modulate the immune system, although the fine mechanism of action is still unanswered. IL binding to cellular PM receptors is well known to involve downstream signaling pathways including signaling kinases and phosphatases [780]. The IL-receptor complexes are also functionally enhanced by carbohydrate-binding ILs [777]. For example, the IL-2 lectin activity for specific oligomanno-saccharides has been suggested to be important [781], because IL-2 is a bifunctional protein upon IL-2 recognition with its IL-2β receptor (IL-2Rβ). Therefore, the N-glycosylated CD3 in the CD3-TCR complex can act as an IL-2 ligand, allowing the downstream phosphorylation of the IL-2Rβ by the downstream pathway of the CD3-TCR-assembled kinase p56lck [781]. This event is also the initial step during the antigen-specific CD4+ T cells activation in immune activation. Upon carbohydrate binding, the accumulated oligomannosaccharides can alter such IL-2 role [782]. In the carbohydrate-binding function of the IL-2-IL-2R complex with glycolipid ligands of adjacent cellular surface complex [781], different ILs target the same receptor. For example, IL-1α and IL-1β target the identical common receptors, while inducing the distinct downstream signaling in each specific target cells. IL-1β, not IL-1α, stimulates human astrocytes, which control the nervous fever generation during inflammatory reactions [783], because astrocytes solely produce IL-1β only with IL-1Rs [784] and GM4 [785] in the central nervous system.

The carbohydrate ligands inhibit the phosphorylation and dephosphorylation induced by ILs. For IL-4 ligand, the NeuAc 1,7 lactone as the carbohydrate ligand is also known on glycoproteins of the human lymphocyte PM [780]. Gangliosides are also reported to modulate IL-4-mediated regulation of the mouse Th-cell line HT-2 [786], acting as a potent suppressor of IL-4-dependent lymphocyte functions through direct recognition of IL-4 and this binding prevents IL-4-IL-4R interaction. In the case of IL-6, IL-6-binding carbohydrate ligand blocks the dephosphorylation of some Tyr-phosphorylation of proteins when stimulated by IL-6 in resting lymphocytes [776]. IL-6 interacts with N-linked HNK-1 carbohydrate epitope of the HSO_3(3-)GlcAβ1,3Galβ1,4GlcNAcβ1,2 carbohydrate sequence [787]. The SO_3H (3-)GlcA moiety is required for the binding to IL-6. Together with the SO_3H (3-) GlcA determinant, the glycan binding motif of IL-6 also contains a methyl group-carrying glycan structure in the 2-acetamido group of GlcNAc or the Fuc methyl group, the HSO_3(3-)GlcAβ1,3Galβ1,4. The IL-6 recognition domain present in the SO_3H(3-)GlcAβ1,3Galβ1,4 sequence, followed by a monosaccharide residue bearing a methyl group, which is either generated from the acetamido group of GlcNAc present in N-glycans or from the Fuc methyl group in oligosaccharide alditols. IL-7 also preferentially binds to the ovine submaxillary mucin. IL-7 can bind to fetuin and mucins, but not to the glycosaminoglycan. SA residue is required for the binding to IL-7; however, the IL-7 binding is not inhibited by Neu5Ac or Neu5Gc or by the two SA mixture. GalNAc-OH, not GalNAc residue, inhibits IL-7 recognition. Another ligand of IL-7 is sialyl-Tn antigen [788].

6.8 Conclusion

Carbohydrates ubiquitously expressed exhibit various functions of protein folding, trafficking, signaling, interaction, embryogenesis, pathogen recognition, and immune responses. Post-translational and co-translational glycosylation events in modification of proteins, lipids, and other carbohydrates are influent for cell adhesion, migration, invasion, signal transduction, and cycle. These phenomena are frequently observed in in vitro and in vivo malignantly transformed cells. The carbohydrate structure is complex and the complexity is further amplified by the stereomeric-isomers, anomeric configurations, antennary branched side chains, and hydroxyl group modifications through sulfation, methylation, and phosphorylation. From the heterogeneity compared to proteomics and genomics, glycomics is faced with the extremely restricted approaches to synthesize glycan structures and analyze functions. The glycan structure complexity is also faced with limited approaches in data analysis. In the PCI and carbohydrate-sphingolipid interaction (CSI), carbohydrate synthesis, analysis, and binding protein analysis are also limited.

The functional interactions of carbohydrates with their receptor proteins or PCI are the evolutionary conserved biological events and the basic pathology. GSL-protein interactions lead to precise biological functions. Upon binding with their interacting proteins, GSLs modulate their activities, requiring deciphering the GSL-glycocode. PCI as protein-GSL carbohydrates interaction has become the great interest in the mechanism elucidation on the biological diversity, evolution, infectious agents' propagation, and therapeutic glycan biomarkers. GSLs exert their distinct functions through binding to certain functional proteins. GSLs influence on receptor signaling on the plasma membrane. The GSLs regulate RTKs, depending on changes in GSLs composition due to the activation and inactivation of the receptors. GSLs function as players in allosteric regulation in receptor functions and consequently contribute to the receptor multimerization and phosphorylation.

The number of proteins interacting with GSLs is presently limited. CCI around the key interaction is also receiving a great attention for the biological meanings. Although GSL-recognizing domains recognize and bind GSL-glycan residues, the chemical and ion bonds are not mechanistically explained. Because glycan carbohydrate residues in GSLs are regarded as a biological "face" or "marker" surfaced on cells to specify their appearances, the information contained in GSL "code" should be decoded or solved, giving the GSL-glycocode. GSL-recognizing domains physically interact with GSL carbohydrate residues by PCI and CCI in order to regulate the downstream signaling. Such cases of GSL-recognizing protein domains have been seen in the viral and bacterial GSL-recognizing domains as well as in the several cellular membrane growth factor receptors, tetraspanins, integrins, and caveolins. Hypoxia alters glycosylation and glycan-receptor interactions such as PCI or CCI. For example, intracellular shift of metabolic pathway of glucose from mitochondrial respiratory pathway to anaerobic cytosol glycolytic pathway leads to retarded integrin $3\alpha1\beta$ protein translocation to the PM, $\alpha1,2$-fucosylation, and galectin overexpression [789]. However, how many GSL-recognizing domains exist is the main question. In addition, how such carbohydrate-binding domains regulate the receptor signals in the cells upon accounting the GSLs? Although GSLs function over the typical ligand-receptor interaction, behaviors of GSLs-specific glycosyltransferases, sub-ER-Golgi network compartmentalization, and localization remain unknown [790].

References

1. Kopitz J (2017) Lipid glycosylation: a primer for histochemists and cell biologists. Histochem Cell Biol 147(2):175–198
2. Merrill AH Jr (2011) Sphingolipid and glycosphingolipid metabolic pathways in the era of sphingolipidomics. Chem Rev 111:6387–6422
3. Merrill AH Jr, Carman GM (2015) Introduction to thematic minireview series: novel bioactive sphingolipids. J Biol Chem 290:15362–15364
4. Thudichum JLW (1884) A treatise on the chemical constitution of the brain. Bailliere, Tindall and Cox, London
5. Björkholm P, Ernst AM, Hacke M, Wieland F, Brügger B, von Heijne G (2014) Identification of novel sphingolipid-binding motifs in mammalian membrane proteins. Biochim Biophys Acta 1838(8):2066–2070
6. Chester MA (1998) IUPAC-IUB joint commission on biochemical nomenclature (JCBN). Nomenclature of glycolipids--recommendations 1997. Eur J Biochem 257(2):293–298
7. IUPAC-IUB Commission on Biochemical Nomenclature (1977) The nomenclature of lipids. (recommendations 1976). Hoppe Seylers Z Physiol Chem 358:617–631
8. Gabius HJ, Roth J (2017) An introduction to the sugar code. Histochem Cell Biol 147(2):111
9. Hakomori S, Ishizuka I (2001) Glycolipid: animal. In Encyclopedia of life sciences. Nature Publishing Group, London
10. Folch J, Lees M, Sloane Stanley GH (1957) A simple method for the isolation and purification of total lipids from animal tissues. J Biol Chem 226:497–509
11. Wong CG, Ladisch S (1983) Retention of gangliosides in serum delipidated by di-isopropyl ether-1-butanol extraction. J Lipid Res 24:666–669
12. Zhang XB, Tang NM (1993) A method for micro-scale ganglioside purification. Acta Biochim Biophys Sinica 25:187–190
13. Saito T, Hakomori SI (1971) Quantitative isolation of total glycosphingolipids from animal cells. J Lipid Res 12:257–279
14. Melton-Celsa AR (2014) Shiga toxin (Stx) classification, structure, and function. Microbiol Spectr 2(4). https://doi.org/10.1128/microbiolspec.EHEC-0024-2013
15. Jennemann R, Gröne HJ (2013) Cell-specific in vivo functions of glycosphingolipids: lessons from genetic deletions of enzymes involved in glycosphingolipid synthesis. Prog Lipid Res 52 (2):231–248
16. Schnaar RL, Kinoshita T (2017) Glycosphingolipids. In: Varki A, Cummings RD, Esko JD, Stanley P, Hart GW, Aebi M, Darvill AG, Kinoshita T, Packer NH, Prestegard JH, Schnaar RL, Seeberger PH (eds) Essentials of glycobiology, 3rd edn. Cold Spring Harbor Laboratory Press, Cold Spring Harbor, NY. 2015-2017. Chapter 11

C.-H. Kim, *Glycosphingolipids Signaling*,
https://doi.org/10.1007/978-981-15-5807-8

17. Russo D, Capolupo L, Loomba JS, Sticco L, D'Angelo G (2018) Glycosphingolipid metabolism in cell fate specification. J Cell Sci 131(24):jcs219204
18. Park SY, Kwak CY, Shayman JA, Kim JH (2012) Globoside promotes activation of ERK by interaction with the epidermal growth factor receptor. Biochim Biophys Acta 1820 (7):1141–1148
19. Guan F, Handa K, Hakomori SI (2011) Regulation of epidermal growth factor receptor through interaction of ganglioside GM3 with GlcNAc of N-linked glycan of the receptor: demonstration in ldlD cells. Neurochem Res 36(9):1645–1653
20. Park HJ, Chae SK, Kim JW, Yang SG, Jung JM, Kim MJ, Wee G, Lee DS, Kim SU, Koo DB (2017) Ganglioside GM3 induces cumulus cell apoptosis through inhibition of epidermal growth factor receptor-mediated PI3K/AKT signaling pathways during in vitro maturation of pig oocytes. Mol Reprod Dev 84(8):702–711
21. Kondo Y, Ikeda K, Tokuda N, Nishitani C, Ohto U, Akashi-Takamura S, Ito Y, Uchikawa M, Kuroki Y, Taguchi R, Miyake K, Zhang Q, Furukawa K, Furukawa K (2013) TLR4-MD-2 complex is negatively regulated by an endogenous ligand, globotetraosylceramide. Proc Natl Acad Sci U S A 110(12):4714–4719
22. Chakrabandhu K, Huault S, Garmy N, Fantini J, Stebe E, Mailfert S, Marguet D, Hueber AO (2008) The extracellular glycosphingolipid-binding motif of Fas defines its internalization route, mode and outcome of signals upon activation by ligand. Cell Death Differ 15 (12):1824–1837
23. Andrews NW, Almeida PE, Corrotte M (2014) Damage control: cellular mechanisms of plasma membrane repair. Trends Cell Biol 24(12):734–742
24. Lindquist S, Karitkina D, Langnaese K, Posevitz-Fejfar A, Schraven B, Xavier R, Seed B, Lindquist JA (2011) Phosphoprotein associated with glycosphingolipid-enriched microdomains differentially modulates SRC kinase activity in brain maturation. PLoS One 6 (9):e23978
25. Basu I, Mukhopadhyay C (2014) Insights into binding of cholera toxin to GM1 containing membrane. Langmuir 30(50):15244–15252
26. Yu X, Coulson BS, Fleming FE, Dyason JC, von Itzstein M, Blanchard H (2011) Novel structural insights into rotavirus recognition of ganglioside glycan receptors. J Mol Biol 413 (5):929–939
27. Karve SS, Weiss AA (2014) Glycolipid binding preferences of Shiga toxin variants. PLoS One 9(7):e101173
28. Surendran-Nair M, Kollanoor-Johny A, Ananda-Baskaran S, Norris C, Lee JY, Venkitanarayanan K (2016) Selenium reduces enterohemorrhagic *Escherichia coli* O157:H7 verotoxin production and globotriaosylceramide receptor expression on host cells. Future Microbiol 11:745–756
29. Brelot A, Chakrabarti LA (2018) CCR5 revisited: how mechanisms of HIV entry govern AIDS pathogenesis. J Mol Biol 430(17):2557–2589
30. De Rosa M, Park HJ, Mylvaganum M et al (2008) The medium is the message: glycosphingolipids and their soluble analogues. Biochim Biophys Acta 1780(3):347–352
31. Low JA, Magnuson B, Tsai B, Imperiale MJ (2006) Identification of gangliosides GD1b and GT1b as receptors for BK virus. J Virol 80(3):1361–1366
32. Ströh LJ, Maginnis MS, Blaum BS, Nelson CD, Neu U, Gee GV, O'Hara BA, Motamedi N, DiMaio D, Atwood WJ, Stehle T (2015) The greater affinity of JC polyomavirus capsid for α2,6-linked lactoseries tetrasaccharide c than for other sialylated glycans is a major determinant of infectivity. J Virol 89(12):6364–6375
33. Tsai B, Gilbert JM, Stehle T, Lencer W, Benjamin TL, Rapoport TA (2003) Gangliosides are receptors for murine polyoma virus and SV40. EMBO J 22(17):4346–4355
34. Erickson KD, Garcea RL, Tsai B (2009) Ganglioside GT1b is a putative host cell receptor for the Merkel cell polyomavirus. J Virol 83(19):10275–10279
35. Brown KE, Anderson SM, Young NS (1993) Erythrocyte P antigen: cellular receptor for B19 parvovirus. Science 262(5130):114–117

36. Brown KE, Liu Z, Gallinella G, Wong S, Mills IP, O'Sullivan MG (2004) Simian parvovirus infection: a potential zoonosis. J Infect Dis 190(11):1900–1907
37. Gorelik L, Reid C, Testa M, Brickelmaier M, Bossolasco S, Pazzi A, Bestetti A, Carmillo P, Wilson E, McAuliffe M, Tonkin C, Carulli JP, Lugovskoy A, Lazzarin A, Sunyaev S, Simon K, Cinque P (2011) Progressive multifocal leukoencephalopathy (PML) development is associated with mutations in JC virus capsid protein VP1 that change its receptor specificity. J Infect Dis 204:103–114
38. Fleming FE, Böhm R, Dang VT, Holloway G, Haselhorst T, Madge PD, Deveryshetty J, Yu X, Blanchard H, von Itzstein M, Coulson BS (2014) Relative roles of GM1 ganglioside, N-acylneuraminic acids, and α2β1 integrin in mediating rotavirus infection. J Virol 88 (8):4558–4571
39. Ströh LJ, Stehle T (2014) Glycan engagement by viruses: receptor switches and specificity. Annu Rev Virol 1(1):285–306
40. Russo D, Parashuraman S, D'Angelo G (2016) Glycosphingolipid-protein interaction in signal transduction. Int J Mol Sci 17(10):pii: E1732
41. Lai CH, Hütter J, Hsu CW, Tanaka H, Varela-Aramburu S, De Cola L, Lepenies B, Seeberger PH (2016) Analysis of carbohydrate-carbohydrate interactions using sugar-functionalized silicon nanoparticles for cell imaging. Nano Lett 16(1):807–811
42. Handa K, Hakomori SI (2011) Carbohydrate to carbohydrate interaction in development process and cancer progression. Glycoconj J 29(8–9):627–637
43. Gabius HJ (2008) Glycans: bioactive signals decoded by lectins. Biochem Soc Trans 36 (Pt 6):1491–1496
44. Haberkant P, Schmitt O, Contreras FX, Thiele C, Hanada K, Sprong H, Reinhard C, Wieland FT, Brügger B (2008) Protein-sphingolipid interactions within cellular membranes. J Lipid Res 49(1):251–262
45. Haberkant P, Stein F, Höglinger D, Gerl MJ, Brügger B, Van Veldhoven PP, Krijgsveld J, Gavin AC, Schultz C (2016) Bifunctional Sphingosine for cell-based analysis of protein-Sphingolipid interactions. ACS Chem Biol 11(1):222–230
46. Lepenies B, Lee J, Sonkaria S (2013) Targeting C-type lectin receptors with multivalent carbohydrate ligands. Adv Drug Deliv Rev 65:1271–1281
47. Lundquist JJ, Toone EJ (2002) The cluster glycoside effect. Chem Rev 102:555–578
48. Boggs JM, Gao W, Hirahara Y (2008) Myelin glycosphingolipids, galactosylceramide and sulfatide, participate in carbohydrate-carbohydrate interactions between apposed membranes and may form glycosynapses between oligodendrocyte and/or myelin membranes. Biochim Biophys Acta Gen Subj 1780:445–455
49. Bucior I, Scheuring S, Engel A, Burger MM (2004) Carbohydrate-carbohydrate interaction provides adhesion force and specificity for cellular recognition. J Cell Biol 165:529–537
50. Kojima N, Fenderson BA, Stroud MR, Goldberg RI, Habermann R, Toyokuni T, Hakomori S (1994) Further studies on cell adhesion based on Le(x)-Le(x) interaction, with new approaches: embryoglycan aggregation of F9 teratocarcinoma cells, and adhesion of various tumour cells based on Le(x) expression. Glycoconj J 11(3):238–248
51. Zhao J, Liu Y, Park HJ, Boggs JM, Basu A (2012) Carbohydrate-coated fluorescent silica nanoparticles as probes for the galactose/3-sulfogalactose carbohydrate-carbohydrate interaction using model systems and cellular binding studies. Bioconjug Chem 23:1166–1173
52. Seah N, Santacroce PV, Basu A. (2009) Probing the lactose. GM3 carbohydrate-carbohydrate
53. Ideo H, Seko A, Ishizuka I, Yamashita K (2003) The N-terminal carbohydrate recognition domain of galectin-8 recognizes specific glycosphingolipids with high affinity. Glycobiology 13(10):713–723
54. Heuss SF, Tarantino N, Fantini J, Ndiaye-Lobry D, Moretti J, Israël A, Logeat F (2013) A glycosphingolipid binding domain controls trafficking and activity of the mammalian notch ligand delta-like 1. PLoS One 8(9):e74392

55. Matsuura K, Oda R, Kitakouji H, Kiso M, Kitajima K, Kobayashi K (2004) Surface plasmon resonance study of carbohydrate-carbohydrate interaction between various gangliosides and Gg3-carrying polystyrene. Biomacromolecules 5:937–941
56. Rhome R, McQuiston T, Kechichian T, Bielawska A, Hennig M, Drago M, Morace G, Luberto C, Del Poeta M (2007) Biosynthesis and immunogenicity of glucosylceramide in Cryptococcus neoformans and other human pathogens. Eukaryot Cell 6(10):1715–1726
57. Bhunia A, Vivekanandan S, Eckert T, Burg-Roderfeld M, Wechselberger R, Romanuka J, Bachle D, Kornilov AV, von der Lieth CW, Jimenez-Barbero J, Nifantiev NE, Schachner M, Sewald N, Lütteke T, Gabius HJ, Siebert HC (2010) Why structurally different cyclic peptides can be glycomimetics of the HNK-1 carbohydrate antigen. J Am Chem Soc 132:96–105
58. Chou DK et al (1986) Structure of sulfated glucuronyl glycolipids in the nervous system reacting with HNK-1 antibody and some IgM paraproteins in neuropathy. J Biol Chem 261:11717–11725
59. Mullen TD, Hannun YA, Obeid LM (2012) Ceramide synthases at the Centre of sphingolipid metabolism and biology. Biochem J 3:789–802
60. Levy M, Futerman AH (2010) Mammalian ceramide synthases. IUBMB Life 62:347–356
61. Wegner MS, Schiffmann S, Parnham MJ, Geisslinger G, Grösch S (2016) The enigma of ceramide synthase regulation in mammalian cells. Prog Lipid Res 63:93–119
62. Koivusalo M, Jansen M, Somerharju P, Ikonen E (2007) Endocytic trafficking of sphingomyelin depends on its acyl chain length. Mol Biol Cell 18:5113–5123
63. Corfield AP (2017) Protein glycosylation: a primer for histochemists and cell biologists. Histochem Cell Biol 147(2):119–147
64. Holthuis JC, Pomorski T, Raggers RJ, Sprong H, Van Meer G (2001) The organizing potential of sphingolipids in intracellular membrane transport. Physiol Rev 81(4):1689–1723
65. Tidhar R, Futerman AH (2013) The complexity of sphingolipid biosynthesis in the endoplasmic reticulum. Biochim Biophys Acta 1833:2511–2518
66. Chiba A, Kusunoki S, Obata H, Machinami R, Kanazawa I (1997) Ganglioside composition of the human cranial nerves, with special reference to pathophysiology of Miller Fisher syndrome. Brain Res 745:32–36
67. Chisada S, Yoshimura Y, Sakaguchi K, Uemura S, Go S, Ikeda K, Uchima H, Matsunaga N, Ogura K, Tai T, Okino N, Taguchi R, Inokuchi J, Ito M (2009) Zebrafish and mouse alpha2,3-sialyltransferases responsible for synthesizing GM4 ganglioside. J Biol Chem 284 (44):30534–30546
68. Hanada K (2017) Ceramide transport from the endoplasmic reticulum to the trans golgi region at organelle membrane contact sites. Adv Exp Med Biol 997:69–81
69. Halter D, Neumann S, van Dijk SM, Wolthoorn J, de Mazière AM, Vieira OV, Mattjus P, Klumperman J, van Meer G, Sprong H (2007) Pre- and post-golgi translocation of glucosylceramide in glycosphingolipid synthesis. J Cell Biol 179:101–115
70. Hanada K, Kumagai K, Yasuda S, Miura Y, Kawano M, Fukasawa M, Nishijima M (2003) Molecular machinery for non-vesicular trafficking of ceramide. Nature 426(6968):803–809
71. Yamaji T, Hanada K (2015) Sphingolipid metabolism and interorganellar transport: localization of sphingolipid enzymes and lipid transfer proteins. Traffic 60:511–518
72. Gillon A, Latham CF, Miller EA (2012) Vesicle-mediated ER export of proteins and lipids. Biochim Biophys Acta 1821:1040–1049
73. Yamaji T, Horie A, Tachida Y, Sakuma C, Suzuki Y, Kushi Y, Hanada K (2016) Role of intracellular lipid logistics in the preferential usage of very long chain-ceramides in glucosylceramide. Int J Mol Sci 17(10):pii: E1761
74. Maccioni HJ, Quiroga R, Ferrari ML (2011) Cellular and molecular biology of glycosphingolipid glycosylation. J Neurochem 117(4):589–602
75. D'Angelo G, Uemura T, Chuang CC, Polishchuk E, Santoro M, Ohvo-Rekilä H, Sato T, Di Tullio G, Varriale A, D'Auria S, Daniele T, Capuani F, Johannes L, Mattjus P, Monti M, Pucci P, Williams RL, Burke JE, Platt FM, Harada A, De Matteis MA (2013) Vesicular and

non-vesicular transport feed distinct glycosylation pathways in the Golgi. Nature 501 (7465):116–120
76. Roth J, Zuber C (2017) Quality control of glycoprotein folding and ERAD: the role of *N*-glycan handling, EDEM1 and OS-9. Histochem Cell Biol 147(2):269–284
77. Kumagai T, Sato T, Natsuka S, Kobayashi Y, Zhou D, Shinkai T, Hayakawa S, Furukawa K (2010) Involvement of murine beta-1,4-galactosyltransferase V in lactosylceramide biosynthesis. Glycoconj J 27(7–9):685–695
78. Smith DC, Sillence DJ, Falguières T, Jarvis RM, Johannes L, Lord JM, Platt FM, Roberts LM (2006) The association of Shiga-like toxin with detergent-resistant membranes is modulated by glucosylceramide and is an essential requirement in the endoplasmic reticulum for a cytotoxic effect. Mol Biol Cell 17(3):1375–1387
79. De Rosa MF, Sillence D, Ackerley C, Lingwood C (2004) Role of multiple drug resistance protein 1 in neutral but not acidic glycosphingolipid biosynthesis. J Biol Chem 279 (9):7867–7876
80. Lingwood CA (2011) Glycosphingolipid functions. Cold Spring Harb Perspect Biol 3(7):pii: a004788
81. Lingwood CA (1996) Role of verotoxin receptors in pathogenesis. Trends Microbiol 4:147–153
82. Wiels J, Fellous M, Tursz T (1981) Monoclonal antibody against a Burkitt lymphoma-associated antigen. Proc Natl Acad Sci U S A 78:6485–6488
83. Distler U, Souady J, Hülsewig M, Drmić-Hofman I, Haier J, Friedrich AW, Karch H, Senninger N, Dreisewerd K, Berkenkamp S, Schmidt MA, Peter-Katalinić J, Müthing J (2009) Shiga toxin receptor Gb3Cer/CD77: tumor-association and promising therapeutic target in pancreas and colon cancer. PLoS One 4:e6813
84. Kovbasnjuk O, Mourtazina R, Baibakov B, Wang T, Elowsky C, Choti MA, Kane A, Donowitz M (2005) The glycosphingolipid globotriaosylceramide in the metastatic transformation of colon cancer. Proc Natl Acad Sci U S A 102:19087–19092
85. Gehrmann M, Liebisch G, Schmitz G, Anderson R, Steinem C, De Maio A, Pockley G, Multhoff G (2008) Tumor-specific Hsp70 plasma membrane localization is enabled by the glycosphingolipid Gb3. PLoS One 3:e1925
86. Clarke JT (2007) Narrative review: Fabry disease. Ann Intern Med 146:425–433
87. Lingwood CA, Binnington B, Manis A, Branch DR (2010) Globotriaosyl ceramide receptor function - where membrane structure and pathology intersect. FEBS Lett 584:1879–1886
88. Siddiqui B, Hakomori S (1973) A ceramide tetrasaccharide of human erythrocyte membrane reacting with anti-type XIV pneumococcal polysaccharide antiserum. Biochim Biophys Acta 330:147–155
89. Willison KR, Karol RA, Suzuki A, Kundu SK, Marcus DM (1982) Neutral glycolipid antigens as developmental markers of mouse teratocarcinoma and early embryos: an immunologic and chemical analysis. J Immunol 129:603–609
90. Yu S, Withers DA, Hakomori S (1998) Globoside-dependent adhesion of human embryonal carcinoma cells, based on carbohydrate-carbohydrate interaction, initiates signal transduction and induces enhanced activity of transcription factors AP1 and CREB. J Biol Chem 273:2517–2525
91. Zeng X, Miura T, Luo Y, Bhattacharya B, Condie B, Chen J, Ginis I, Lyons I, Mejido J, Puri RK, Rao MS, Freed WJ (2004) Properties of pluripotent human embryonic stem cells BG01 and BG02. Stem Cells 22:292–312
92. Zhao Y, Wang H, Mazzone T (2006) Identification of stem cells from human umbilical cord blood with embryonic and hematopoietic characteristics. Exp Cell Res 312:2454–2464
93. Liour SS, Kraemer SA, Dinkins MB, Su CY, Yanagisawa M, Yu RK (2006) Further characterization of embryonic stem cell-derived radial glial cells. Glia 53:43–56
94. Liang YJ, Kuo HH, Lin CH, Chen YY, Yang BC, Cheng YY, Yu AL, Khoo KH, Yu J (2010) Switching of the core structures of glycosphingolipids from globo- and lacto- to ganglio-series upon human embryonic stem cell differentiation. Proc Natl Acad Sci U S A 107:22564–22569

95. Bieberich E (2012) It's a lipid's world: bioactive lipid metabolism and signaling in neural stem cell differentiation. Neurochem Res 37(6):1208–1229
96. Maccioni HJ, Quiroga R, Spessott W (2011) Organization of the synthesis of glycolipid oligosaccharides in the golgi complex. FEBS Lett 585(11):1691–1698
97. Klenk E (1942) Über die ganglioside, eine neue Gruppe von zuckerhaltigen Gehirnlipoiden. Z Physiol Chem 273:76
98. Svennerholm L (1963) Chromatographic separation of human brain gangliosides. J Neurochem 10:613–623
99. Stencel-Baerenwald JE, Reiss K, Reiter DM, Stehle T, Dermody TS (2014) The sweet spot: defining virus-sialic acid interactions. Nat Rev Microbiol 12:739–749
100. Ströh LJ, Stehle T (2014) Glycan engagement by viruses: receptor switches and specificity. Ann Rev Virol 1:285–306
101. Varki A (2001) Loss of N-glycolylneuraminic acid in humans: mechanisms, consequences, and implications for hominid evolution. Am J Phys Anthropol Suppl 33:54–69
102. Daniotti JL (2011) Iglesias-Bartolome, R. metabolic pathways and intracellular trafficking of gangliosides. IUBMB Life 63:513–520
103. Cuervo AM, Stefanis L, Fredenburg R, Lansbury PT, Sulzer D (2004) Impaired degradation of mutant alpha-synuclein by chaperone-mediated autophagy. Science 305(5688):1292–1295
104. Huwiler A, Kolter T, Pfeilschifter J, Sandhoff K (2000) Physiology and pathophysiology of sphingolipid metabolism and signaling. Biochim Biophys Acta 1485(2–3):63–99
105. Zhang X, Kiechle FL (2004) Review: glycosphingolipids in health and disease. Ann Clin Lab Sci 34(1):3–13
106. Ferrari G, Greene LA (1998) Promotion of neuronal survival by GM1 ganglioside. Phenomenology and mechanism of action. Ann N Y Acad Sci 845:263–273
107. Ferrari G, Anderson BL, Stephens RM, Kaplan DR, Greene LA (1995) Prevention of apoptotic neuronal death by GM1 ganglioside. Involvement of Trk neurotrophin receptors. J Biol Chem 270:3074–3080
108. Choi JS, Kim JA, Joo CK (2003) Activation of MAPK and CREB by GM1 induces survival of RGCs in the retina with axotomized nerve. Invest Ophthalmol Vis Sci 44:1747–1752
109. Bremer EG, Schlessinger J, Hakomori S (1986) Ganglioside-mediated modulation of cell growth. Specific effects of GM3 on tyrosine phosphorylation of the epidermal growth factor receptor. J Biol Chem 261(5):2434–2440
110. Hanai N, Nores GA, MacLeod C, Torres-Mendez CR, Hakomori S (1988) Ganglioside-mediated modulation of cell growth. Specific effects of GM3 and lyso-GM3 in tyrosine phosphorylation of the epidermal growth factor receptor. J Biol Chem 263(22):10915–10921
111. Nicholson RI, Gee JMW, Harper ME (2001) EGFR and cancer prognosis. Eur J Cancer 37:9–15
112. Kawashima N, Yoon SJ, Itoh K, Nakayama K (2009) Tyrosine kinase activity of epidermal growth factor receptor is regulated by GM3 binding through carbohydrate to carbohydrate interactions. J Biol Chem 284(10):6147–6155
113. Choi HJ, Chung TW, Kang SK, Lee YC, Ko JH, Kim JG, Kim CH (2006) Ganglioside GM3 modulates tumor suppressor PTEN-mediated cell cycle progression—transcriptional induction of p21(WAF1) and p27(kip1) by inhibition of PI-3K/AKT pathway. Glycobiology 16 (7):573–583
114. Choi HJ, Chung TW, Kim SJ, Cho SY, Lee YS, Lee YC, Ko JH, Kim CH (2008) The AP-2alpha transcription factor is required for the ganglioside GM3-stimulated transcriptional regulation of a PTEN gene. Glycobiology 18(5):395–407
115. Hanai N, Dohi T, Nores GA, Hakomori S (1988) A novel ganglioside, de-N-acetyl-GM3 (II3NeuNH2LacCer), acting as a strong promoter for epidermal growth factor receptor kinase and as a stimulator for cell growth. J Biol Chem 263(13):6296–6301
116. Zurita AR, Maccioni HJ, Daniotti JL (2001) Modulation of epidermal growth factor receptor phosphorylation by endogenously expressed gangliosides. Biochem J 355(Pt 2):465–472

117. Krengel U, Bousquet PA (2014) Molecular recognition of gangliosides and their potential for cancer immunotherapies. Front Immunol 5:325
118. Arkhipov A, Shan Y, Das R, Endres NF, Eastwood MP, Wemmer DE, Kuriyan J, Shaw DE (2013) Architecture and membrane interactions of the EGF receptor. Cell 152(3):557–569
119. Todeschini AR, Dos Santos JN, Handa K, Hakomori S (2008) Ganglioside GM2/GM3 complex affixed on silica nanospheres strongly inhibits cell motility through CD82/ cMet-mediated pathway. Proc Natl Acad Sci U S A 105:1925–1930
120. Todeschini AR, Dos Santos JN, Handa K, Hakomori S (2007) Ganglioside GM2-tetraspanin CD82 through glycosynapse. J Biol Chem 282:8123–8133
121. Liang YJ, Wang CY, Wang IA, Chen YW, Li LT, Lin CY, Ho MY, Chou TL, Wang YH, Chiou SP, Lin YJ, Yu J (2017) Interaction of glycosphingolipids GD3 and GD2 with growth factor receptors maintains breast cancer stem cell phenotype. Oncotarget 8(29):47454–47473
122. Liang YJ, Ding Y, Levery SB, Lobaton M, Handa K, Hakomori S (2013) Differential expression profiles of glycosphingolipids in human breast cancer stem cells vs. cancer non-stem cells. Proc Natl Acad Sci U S A 110:4968–4973
123. Battula VL, Shi Y, Evans KW, Wang RY, Spaeth EL, Jacamo RO, Guerra R, Sahin AA, Marini FC, Hortobagyi G, Mani SA, Andreeff M (2012) Ganglioside GD2 identifies breast cancer stem cells and promotes tumorigenesis. J Clin Invest 122:2066–2078
124. Liu Y, Li R, Ladisch S (2004) Exogenous ganglioside GD1a enhances epidermal growth factor receptor binding and dimerization. J Biol Chem 279(35):36481–36489
125. Mirkin BL, Clark SH, Zhang C (2002) Inhibition of human neuroblastoma cell proliferation and EGF receptor phosphorylation by gangliosides GM1, GM3, GD1A and GT1B. Cell Prolif 35(2):105–115
126. Toledo MS, Suzuki E, Handa K, Hakomori S (2005) Effect of ganglioside and tetraspanins in microdomains on interaction of integrins with fibroblast growth factor receptor. J Biol Chem 280(16):16227–16234
127. Hakomori SI, Handa K (2015) GM3 and cancer. Glycoconj J 32(1–2):1–8
128. Schnaar RL (2010) Brain gangliosides in axon-myelin stability and axon regeneration. FEBS Lett 584(9):1741–1747
129. Simpson MA, Cross H, Proukakis C, Priestman DA, Neville DC, Reinkensmeier G, Wang H, Wiznitzer M, Gurtz K, Verganelaki A, Pryde A, Patton MA, Dwek RA, Butters TD, Platt FM, Crosby AH (2004) Infantile-onset symptomatic epilepsy syndrome caused by a homozygous loss-of-function mutation of GM3 synthase. Nat Genet 36(11):1225–1229
130. Harlalka GV, Lehman A, Chioza B, Baple EL, Maroofian R, Cross H, Sreekantan-Nair A, Priestman DA, Al-Turki S, McEntagart ME, Proukakis C, Royle L, Kozak RP, Bastaki L, Patton M, Wagner K, Coblentz R, Price J, Mezei M, Schlade-Bartusiak K, Platt FM, Hurles ME, Crosby AH (2013) Mutations in B4GALNT1 (GM2 synthase) underlie a new disorder of ganglioside biosynthesis. Brain 136(Pt 12):3618–3624
131. Kojima N, Hakomori S (1989) Specific interaction between gangliotriaosylceramide (Gg3) and sialosyllactosylceramide (G(M3)) as a basis for specific cellular recognition between lymphoma and melanoma cells. J Biol Chem 264(34):20159–20162
132. Ichikawa N, Iwabuchi K, Kurihara H, Ishii K, Kobayashi T, Sasaki T, Hattori N, Mizuno Y, Hozumi K, Yamada Y, Arikawa-Hirasawa E (2009) Binding of laminin-1 to monosialoganglioside GM1 in lipid rafts is crucial for neurite outgrowth. J Cell Sci 122 (Pt 2):289–299
133. Mutoh T, Hamano T, Tokuda A, Kuriyama M (2000) Unglycosylated Trk protein does not co-localize nor associate with ganglioside GM1 in stable clone of PC12 cells overexpressing Trk (PCtrk cells). Glycoconj J 17(3–4):233–237
134. Ledeen RW, Wu G (2015) The multi-tasked life of GM1 ganglioside, a true factotum of nature. Trends Biochem Sci 40(7):407–418
135. Wu, et al. (2009) Sodium-calcium exchanger complexed with GM1 ganglioside in nuclear membrane transfers calcium from nucleoplasm to endoplasmic reticulum. Proc Natl Acad Sci U S A 106:10829–10834

136. Fukuda Y, Fukui T, Hikichi C, Ishikawa T, Murate K, Adachi T, Imai H, Fukuhara K, Ueda A, Kaplan AP, Mutoh T (2015) Neurotropin promotes NGF signaling through interaction of GM1 ganglioside with Trk neurotrophin receptor in PC12 cells. Brain Res 1596:13–21
137. Longo FM, Massa SM (2013) Small-molecule modulation of neurotrophin receptors: a strategy for the treatment of neurological disease. Nat Rev 12:507–525
138. Scarpi D, Cirelli D, Matrone C, Castronovo G, Rosini P, Occhiato EG, Romano F, Bartali L, Clemente AM, Bottegoni G, Cavalli A, De Chiara D, Bonini P, Calissano P, Palamara AT, Garaci E, Torcia MG, Guarna A, Cozzolino F (2012) Low molecular weight, non-peptidic agonists of TrkA receptor with NGF-mimetic activity. Cell Death Dis 3:e339
139. Di Pasquale E, Fantini J, Chahinian H, Maresca M, Taïeb N, Yahi N (2010) Altered ion channel formation by the Parkinson's-disease-linked E46K mutant of alpha-synuclein is corrected by GM3 but not by GM1 gangliosides. J Mol Biol 397(1):202–218
140. Huang F, Dong X, Zhang L, Zhang X, Zhao D, Bai X, Li Z (2009) The neuroprotective effects of NGF combined with GM1 on injured spinal cord neurons in vitro. Brain Res Bull 79 (1):85–88
141. Farooqui T, Franklin T, Pearl DK, Yates AJ (1997) Ganglioside GM1 enhances induction by nerve growth factor of a putative dimer of TrkA. J Neurochem 68:2348–2355
142. Chiricozzi E, Pomè DY, Maggioni M, Di Biase E, Parravicini C, Palazzolo L, Loberto N, Eberini I, Sonnino S (2017) Role of the GM1 ganglioside oligosaccharide portion in the TrkA-dependent neurite sprouting in neuroblastoma cells. J Neurochem 143(6):645–659
143. Zakharova IO, Sokolova TV, Vlasova YA, Furaev VV, Rychkova MP, Avrova NF (2014) GM1 ganglioside activates ERK1/2 and Akt downstream of Trk tyrosine kinase and protects PC12 cells against hydrogen peroxide toxicity. Neurochem Res 39(11):2262–2275
144. Forsayeth J, Hadaczek P (2018) Ganglioside metabolism and Parkinson's disease. Front Neurosci 12:45
145. Zhao J, Singleton PA, Brown ME, Dudek SM, Garcia JG (2009) Phosphotyrosine protein dynamics in cell membrane rafts of sphingosine-1-phosphate-stimulated human endothelium: role in barrier enhancement. Cell Signal 21(12):1945–1960
146. Fantini J, Yahi N (2011) Molecular basis for the glycosphingolipid-binding specificity of α-synuclein: key role of tyrosine 39 in membrane insertion. J Mol Biol 408:654–669
147. Yahi N, Fantini J (2014) Deciphering the glycolipid code of Alzheimer's and Parkinson's amyloid proteins allowed the creation of a universal ganglioside-binding peptide. PLoS One 9: e104751
148. Gavella M, Kveder M, Lipovac V (2010) Modulation of ROS production in human leukocytes by ganglioside micelles. Braz J Med Biol Res 43:942–949
149. Kanatsu Y, Chen NH, Mitoma J, Nakagawa T, Hirabayashi Y, Higashi H (2012) Gangliosides stimulate bradykinin B2 receptors to promote calmodulin kinase II-mediated neuronal differentiation. J Biochem 152(1):63–72
150. Chen N, Furuya S, Doi H, Hashimoto Y, Kudo Y, Higashi H (2003) Ganglioside/calmodulin kinase II signal inducing cdc42-mediated neuronal actin reorganization. Neuroscience 120:163–176
151. Hudmon A, Schulman H (2002) Neuronal CA2+/calmodulin-dependent protein kinase II: the role of structure and autoregulation in cellular function. Annu Rev Biochem 71:473–510
152. Bayer KU, Löhler J, Schulman H, Harbers K (1999) Developmental expression of the CaM kinase II isoforms: ubiquitous γ- and δ-CaM kinase II are the early isoforms and most abundant in the developing nervous system. Brain Res Mol Brain Res 70:147–154
153. Mizumura K, Sugiura T, Katanosaka K, Banik RK, Kozaki Y (2009) Excitation and sensitization of nociceptors by bradykinin: what do we know? Exp Brain Res 196:53–65
154. Shimazaki A, Nakagawa T, Mitoma J, Higashi H (2012) Gangliosides and chondroitin sulfate desensitize and internalize B2 bradykinin receptors. Biochem Biophys Res Commun 420 (1):193–198

155. Tsuji S, Yamashita T, Matsuda Y, Nagai Y (1992) A novel glycosignaling system: GQ1b-dependent neuritogenesis of human neuroblastoma cell line, GOTO, is closely associated with GQ1b-dependent ecto-type protein phosphorylation. Neurochem Int 21:549–554
156. Bliss TVP, Collingridge GL (1993) A synaptic model of memory: long-term potentiation in the hippocampus. Nature 361:31–39
157. Ben-Ari Y, Aniksztejn L, Bregestovski P (1992) Protein kinase C modulation of NMDA currents: an important link for LTP induction. Trends Neurosci 5:333–339
158. Huang YH, Lin Y, Brown TE, Han MH, Saal DB, Neve RL, Zukin RS, Sorg BA, Nestler EJ, Malenka RC, Dong Y (2008) CREB modulates the functional output of nucleus accumbens neurons: a critical role of N-methyl-D-aspartate glutamate receptor (NMDAR) receptors. J Biol Chem 283:2751–2760
159. Pascual A et al (2008) Absolute requirement of GDNF for adult catecholaminergic neuron survival. Nat Neurosci 11:755–761
160. Tansey MG et al (2000) GFRα-mediated localization of RET to lipid rafts is required for effective downstream signaling, differentiation, and neuronal survival. Neuron 25:611–623
161. Hadaczek P et al (2015) GDNF signaling implemented by GM1 ganglioside; failure in Parkinson's disease and GM1-deficient murine model. Exp Neurol 263(2015):177–188
162. Hadjiconstantinou M et al (1989) GM1 ganglioside-induced recovery of nigrostriatal dopaminergic neurons after MPTP: an immunohistochemical study. Brain Res 484(1989):297–303
163. Schneider JS et al (1995) GM1 ganglioside rescues substantia nigra pars compacta neurons and increases dopamine synthesis in residual nigrostriatal dopaminergic neurons in MPTP-treated mice. J Neurosci Res 42(1995):117–123
164. Schneider JS, DiStefano L (1995) Response of the damaged dopamine system to GM1 and semisynthetic gangliosides: effects of dose and extent of lesion. Neuropharmacology 34 (1995):489–493
165. Schneider JS et al (2013) A randomized, controlled, delayed start trial of GM1 ganglioside in treated Parkinson's disease patients. J Neurol Sci 324(2013):140–148
166. Wu G et al (2011) Mice lacking major brain gangliosides develop Parkinsonism. Neurochem Res 36:1706–1714
167. Wu G et al (2012) Deficiency of ganglioside GM1 correlates with Parkinson's disease in mice and humans. J Neurosci Res 90:1997–2008
168. Martinez Z et al (2007) GM1 specifically interacts with α-synuclein and inhibits fibrillation. Biochemistry 46:1868–1877
169. Bartels T et al (2014) N-alpha-acetylation of α-synuclein increases its helical folding propensity, GM1 binding specificity and resistance to aggregation. PLoS One 9:e103727
170. Heldt SA, Stanek L, Chhatwal JP, Ressler KJ (2007) Hippocampus-specific deletion of BDNF in adult mice impairs spatial memory and extinction of aversive memories. Mol Psychiatry 12:656–670
171. Inokuchi J, Mizutani A, Jimbo M, Usuki S, Yamagishi K, Mochizuki H, Muramoto K, Kobayashi K, Kuroda Y, Iwasaki K, Ohgami Y, Fujiwara M (1997) Up-regulation of ganglioside biosynthesis, functional synapse formation, and memory retention by a synthetic ceramide analog (l-PDMP). Biochem Biophys Res Commun 237:595–600
172. Fujii S, Igarashi K, Sasaki H, Furuse H, Ito K, Kaneko K, Kato H, Inokuchi J, Waki H, Ando S (2002) Effects of the mono- and tetrasialogangliosides GM1 and GQ1b on ATP-induced long-term potentiation in hippocampal CA1 neurons. Glycobiology 12(5):339–344
173. Shin MK, Jung WR, Kim HG, Roh SE, Kwak CH, Kim CH, Kim SJ, Kim KL (2014) The ganglioside GQ1b regulates BDNF expression via the NMDA receptor signaling pathway. Neuropharmacology 77:414–421
174. Jung WR, Kim HG, Kim KL (2008) Ganglioside GQ1b improves spatial learning and memory of rats as measured by the Y-maze and the Morris water maze tests. Neurosci Lett 439 (2):220–225

175. Shin MK, Choi MS, Chae HJ, Kim JW, Kim HG, Kim KL (2019) Ganglioside GQ1b ameliorates cognitive impairments in an Alzheimer's disease mouse model, and causes reduction of amyloid precursor protein. Sci Rep 9(1):8512
176. Barrier L et al (2007) Genotype-related changes of ganglioside composition in brain regions of transgenic mouse models of Alzheimer's disease. Neurobiol Aging 28:1863–1872
177. Gottfries C-G, Karlsson I, Svennerholm L (1996) Membrane components separate early-onset Alzheimer's disease from senile dementia of the Alzheimer type. Int Psychogeriatr 8:365–372
178. Kracun I, Kalanj S, Talan-Hranilovic J, Cosovic C (1992) Cortical distribution of gangliosides in Alzheimer's disease. Neurochem Int 20:433–438
179. Svennerholm L, Gottfries CG (1994) Membrane lipids, selectively diminished in Alzheimer brains, suggest synapse loss as a primary event in early-onset form (type I) and demyelination in late-onset form (type II). J Neurochem 62:1039–1047
180. Yanagisawa K (2005) GM1 ganglioside and the seeding of amyloid in Alzheimer's disease: endogenous Seed for Alzheimer amyloid. Neuroscientist 11:250–260
181. Okada T, Wakabayashi M, Ikeda K, Matsuzaki K (2007) Formation of toxic fibrils of Alzheimer's amyloid β-protein-(1–40) by monosialoganglioside GM1, a neuronal membrane component. J Mol Biol 371:481–489
182. Ariga T, Wakade C, Yu RK (2011) The pathological roles of Ganglioside metabolism in Alzheimer's disease: effects of gangliosides on neurogenesis. Int J Alzheimers Dis:14
183. Oikawa N et al (2009) Gangliosides determine the amyloid pathology of Alzheimer's disease. Neuroreport 20:1043–1046
184. Nishinaka T, Iwata D, Shimada S, Kosaka K, Suzuki Y (1993) Anti-ganglioside GD1a monoclonal antibody recognizes senile plaques in the brains of patients with Alzheimer-type dementia. Neurosci Res 17:171–176
185. Schengrund C-L (1990) The role (s) of gangliosides in neural differentiation and repair: a perspective. Brain Res Bull 24:131–141
186. Newburn EN, Duchemin AM, Neff NH, Hadjiconstantinou M (2014) GM1 ganglioside enhances Ret signaling in striatum. J Neurochem 130(4):541–554
187. Cazet A, Lefebvre J, Adriaenssens E, Julien S, Bobowski M, Grigoriadis A, Tutt A, Tulasne D, Le Bourhis X, Delannoy P (2010) GD(3) synthase expression enhances proliferation and tumor growth of MDA-MB-231 breast cancer cells through c-Met activation. Mol Cancer Res 8(11):1526–1535
188. Cazet A, Groux-Degroote S, Teylaert B, Kwon KM, Lehoux S, Slomianny C, Kim CH, Le Bourhis X, Delannoy P (2009) GD3 synthase overexpression enhances proliferation and migration of MDA-MB-231 breast cancer cells. Biol Chem 390(7):601–609
189. Cazet A, Bobowski M, Rombouts Y, Lefebvre J, Steenackers A, Popa I, Guérardel Y, Le Bourhis X, Tulasne D, Delannoy P (2012) The ganglioside G(D2) induces the constitutive activation of c-Met in MDA-MB-231 breast cancer cells expressing the G(D3) synthase. Glycobiology 22(6):806–816
190. Li Y, Huang X, Zhong W, Zhang J, Ma K (2013) Ganglioside GM3 promotes HGF-stimulated motility of murine hepatoma cell through enhanced phosphorylation of cMet at specific tyrosine sites and PI3K/Akt-mediated migration signaling. Mol Cell Biochem 382(1–2):83–92
191. Sarkar TR, Battula VL, Werden SJ, Vijay GV, Ramirez-Peña EQ, Taube JH, Chang JT, Miura N, Porter W, Sphyris N, Andreeff M, Mani SA (2015) GD3 synthase regulates epithelial-mesenchymal transition and metastasis in breast cancer. Oncogene 34 (23):2958–2967
192. Li Y, Huang X, Wang C, Li Y, Luan M, Ma K (2015) Ganglioside GM3 exerts opposite effects on motility via epidermal growth factor receptor and hepatocyte growth factor receptor-mediated migration signaling. Mol Med Rep 11(4):2959–2966
193. Hyuga S, Kawasaki N, Hyuga M, Ohta M, Shibayama R, Kawanishi T, Yamagata S, Yamagata T, Hayakawa T (2001) Ganglioside GD1a inhibits HGF-induced motility and scattering of cancer cells through suppression of tyrosine phosphorylation of c-Met. Int J Cancer 94(3):328–334

194. Hynds DL, Summers M, Van Brocklyn J, O'Dorisio MS, Yates AJ (1995) Gangliosides inhibit platelet-derived growth factor-stimulated growth, receptor phosphorylation, and dimerization in neuroblastoma SH-SY5Y cells. J Neurochem 65(5):2251–2258
195. Ohkawa Y, Momota H, Kato A, Hashimoto N, Tsuda Y, Kotani N, Honke K, Suzumura A, Furukawa K, Ohmi Y, Natsume A, Wakabayashi T, Furukawa K (2015) Ganglioside GD3 enhances invasiveness of Gliomas by forming a complex with platelet-derived growth factor receptor α and yes kinase. J Biol Chem 290(26):16043–16058
196. Shih AH, Holland EC (2006) Platelet-derived growth factor (PDGF) and glial tumorigenesis. Cancer Lett 232:139–147
197. MacDonald TJ, Brown KM, LaFleur B, Peterson K, Lawlor C, Chen Y, Packer RJ, Cogen P, Stephan DA (2001) Expression profiling of medulloblastoma: PDGFRA and the RAS/MAPK pathway as therapeutic targets for metastatic disease. Nat Genet 29:143–152
198. Sachinidis A, Kraus R, Seul C, Meyer zu Brickwedde MK, Schulte K, Ko Y, Hoppe J, Vetter H (1996) Gangliosides GM1, GM2 and GM3 inhibit the platelet-derived growth factor-induced signalling transduction pathway in vascular smooth muscle cells by different mechanisms. Eur J Cell Biol 71(1):79–88
199. Oblinger JL, Boardman CL, Yates AJ, Burry RW (2003) Domain-dependent modulation of PDGFRbeta by ganglioside GM1. J Mol Neurosci 20(2):103–114
200. Farooqui T, Kelley T, Coggeshall KM, Rampersaud AA, Yates AJ (1999) GM1 inhibits early signaling events mediated by PDGF receptor in cultured human glioma cells. Anticancer Res 19:5007–5013
201. Lang Z, Guerrera M, Li R, Ladisch S (2001) Ganglioside GD1a enhances VEGF-induced endothelial cell proliferation and migration. Biochem Biophys Res Commun 282 (4):1031–1037
202. Chung TW, Kim SJ, Choi HJ, Kim KJ, Kim MJ, Kim SH, Lee HJ, Ko JH, Lee YC, Suzuki A, Kim CH (2009) Ganglioside GM3 inhibits VEGF/VEGFR-2-mediated angiogenesis: direct interaction of GM3 with VEGFR-2. Glycobiology 19(3):229–239
203. Chung TW, Kim EY, Kim SJ, Choi HJ, Jang SB, Kim KJ, Ha SH, Abekura F, Kwak CH, Kim CH, Ha KT (2017) Sialyllactose suppresses angiogenesis by inhibiting VEGFR-2 activation, and tumor progression. Oncotarget 8(35):58152–58162
204. Coskun Ü, Grzybek M, Drechsel D, Simons K (2011) Regulation of human EGF receptor by lipids. Proc Natl Acad Sci U S A 108(22):9044–9048
205. Mukherjee P, Faber AC, Shelton LM, Baek RC, Chiles TC, Seyfried TN (2008) *Thematic Review Series: Sphingolipids*. Ganglioside GM3 suppresses the proangiogenic effects of vascular endothelial growth factor and ganglioside GD1a. J Lipid Res 49(5):929–938
206. Jin J, Sison K, Li C, Tian R, Wnuk M, Sung HK, Jeansson M, Zhang C, Tucholska M, Jones N, Kerjaschki D, Shibuya M, Fantus IG, Nagy A, Gerber HP, Ferrara N, Pawson T, Quaggin SE (2012) Soluble FLT1 binds lipid microdomains in podocytes to control cell morphology and glomerular barrier function. Cell 151(2):384–399
207. Seyfried TN, Mukherjee P (2010) Ganglioside GM3 is Antiangiogenic in malignant brain cancer. J Oncol 2010:961243
208. Alessandri G, Cornaglia-Ferraris P, Gullino PM (1997) Angiogenic and angiostatic microenvironment in tumors: role of gangliosides. Acta Oncol 36(4):383–387
209. Lewis CE, Pollard JW (2006) Distinct role of macrophages in different tumor microenvironments. Cancer Res 66:605–612
210. Ruffell B, Affara NI, Coussens LM (2012) Differential macrophage programming in the tumor microenvironment. Trends Immunol 33:119–126
211. Liu Y, McCarthy J, Ladisch S (2006) Membrane ganglioside enrichment lowers the threshold for vascular endothelial cell angiogenic signaling. Cancer Res 66:10408–10414
212. Liu Y, Wondimu A, Yan S, Bobb D, Ladisch S (2014) Tumor gangliosides accelerate murine tumor angiogenesis. Angiogenesis 17:563–571

213. Ecsedy JA, Yohe HC, Bergeron AJ, Seyfried TN (1998) Tumor-infiltrating macrophages influence the glycosphingolipid composition of murine brain tumors. J Lipid Res 39:2218–2227
214. Kim SJ, Chung TW, Choi HJ, Jin UH, Ha KT, Lee YC, Kim CH (2014) Monosialic ganglioside GM3 specifically suppresses the monocyte adhesion to endothelial cells for inflammation. Int J Biochem Cell Biol 46:32–38
215. Liu Y, Yan S, Wondimu A, Bob D, Weiss M, Sliwinski K, Villar J, Notario V, Sutherland M, Colberg-Poley AM, Ladisch S (2010) Ganglioside synthase knockout in oncogene-transformed fibroblasts depletes gangliosides and impairs tumor growth. Oncogene 29:3297–3306
216. Wang H, Isaji T, Satoh M, Li D, Arai Y, Gu J (2013) Antitumor effects of exogenous ganglioside gm3 on bladder cancer in an orthotopic cancer model. Urology 81:210 e211–210 e215
217. Ponimaskin E, Dityateva G, Ruonala MO, Fukata M, Fukata Y, Kobe F, Wouters FS, Delling M, Bredt DS, Schachner M, Dityatev A (2008) Fibroblast growth factor-regulated palmitoylation of the neural cell adhesion molecule determines neuronal morphogenesis. J Neurosci 28(36):8897–8907
218. Basilico C, Moscatelli D (1992) The FGF family of growth factors and oncogenes. Adv Cancer Res 59:115–165
219. Mason I (2007) Initiation to end point: the multiple roles of fibroblast growth factors in neural development. Nat Rev Neurosci 8(8):583–596
220. Hynds DL, Burry RW, Yates AJ (1997) Gangliosides inhibit growth factor-stimulated neurite outgrowth in SH-SY5Y human neuroblastoma cells. J Neurosci Res 47:617–625
221. Li R, Manela J, Kong Y, Ladish S (2000) Cellular gangliosides promote growth factor-induced proliferation of fibroblasts. J Biol Chem 275:34213–34223
222. Boldin SA, Futerman AH (2000) Up-regulation of glucosylceramide synthesis upon stimulation of axonal growth by basic fibroblast growth factor. . Evidence for post-translational modification of glucosylceramide synthase. J Biol Chem 275:9905–9909
223. Besancon F, Ankel H (1974) Binding of interferon to gangliosides. Nature (London) 252:478–480
224. Rusnati M, Urbinati C, Tanghetti E, Dell'Era P, Lortat-Jacob H, Presta M (2002) Cell membrane GM1 ganglioside is a functional coreceptor for fibroblast growth factor 2. Proc Natl Acad Sci U S A 99:4367–4372
225. Rusnati M, Tanghetti E, Urbinati C, Tulipano G, Marchesini S, Ziche M, Presta M (1999) Interaction of fibroblast growth factor-2 (FGF-2) with free gangliosides: biochemical characterization and biological consequences in endothelial cell cultures. Mol Biol Cell 10:313–327
226. Wang Q, Song YH, Tang Z, Wang ZP, Xu Q, Bao N (2016) Effects of ganglioside GM1 and neural growth factor on neural stem cell proliferation and differentiation. Genet Mol Res 15(3)
227. Zhang H, Wang JZ, Sun HY, Zhang JN, Yang SY (2004) The effects of GM1 and bFGF synergistically inducing adult rat bone marrow stromal cells to form neural progenitor cells and their differentiation. Chin J Traumatol 7(1):3–6
228. Slevin M, Kumar S, He X, Gaffney J (1999) Physiological concentrations of gangliosides GM1, GM2 and GM3 differentially modify basic-fibroblast-growth-factor-induced mitogenesis and the associated signalling pathway in endothelial cells. Int J Cancer 82 (3):412–423
229. Tzeng SF, Deibler GE, DeVries GH (1999) Myelin basic protein and myelin basic protein peptides induce the proliferation of Schwann cells via ganglioside GM1 and the FGF receptor. Neurochem Res 24(2):255–260
230. Meuillet E, Crémel G, Hicks D, Dreyfus H (1996) Ganglioside effects on basic fibroblast and epidermal growth factor receptors in retinal glial cells. J Lipid Mediat Cell Signal 14 (1–3):277–288
231. Thiery JP (2003) Epithelial–mesenchymal transitions in development and pathologies. Curr Opin Cell Biol 15:740–746

232. Kim SJ, Chung TW, Choi HJ, Kwak CH, Song KH, Suh SJ, Kwon KM, Chang YC, Park YG, Chang HW, Kim KS, Kim CH, Lee YC (2013) Ganglioside GM3 participates in the TGF-β1-induced epithelial-mesenchymal transition of human lens epithelial cells. Biochem J 449 (1):241–251
233. Kwak DH, Lee S, Kim SJ, Ahn SH, Song JH, Choo YK, Choi BK, Jung KY (2005) Ganglioside GM3 inhibits the high glucose- and TGF-beta1-induced proliferation of rat glomerular mesangial cells. Life Sci 77(20):2540–2551
234. Guan F, Handa K, Hakomori SI (2009) Specific glycosphingolipids mediate epithelial-to-mesenchymal transition of human and mouse epithelial cell lines. Proc Natl Acad Sci U S A 106:7461–7466
235. Guan F, Schaffer L, Handa K, Hakomori SI (2010) Functional role of gangliotetraosylceramide in epithelial-to-mesenchymal transition process induced by hypoxia and by TGF-(beta). FASEB J 24:4889–4903
236. Miyazaki H, Fukumoto S, Okada M, Hasegawa T, Furukawa K, Furukawa K (1997) Expression cloning of rat cDNA encoding UDP-galactose:GD2 β1,3-galactosyltransferase that determines the expression of GD1b/GM1/GA1. J Biol Chem 272:24794–24799
237. Giraudo CG, Daniotti JL, Maccioni HJ (2001) Physical and functional association of glycolipid N-acetyl-galactosaminyl and galactosyl transferases in the Golgi apparatus. Proc Natl Acad Sci U S A 98:1625–1630
238. Inman GJ, Nicolas FJ, Hill CS (2002) Nucleocytoplasmic shuttling of Smads 2, 3, and 4 permits sensing of TGF-beta receptor activity. Mol Cell 10:283–294
239. Peinado H, Portillo F, Cano A (2004) Transcriptional regulation of cadherins during development and carcinogenesis. Int J Dev Biol 48:365–375
240. Guo J, Song B, Li X, Hε C, Yang G, Yang X, Guan F (2015) Downregulation of gangliotetraosylceramide and β1,3-galactosyltransferase-4 gene expression by Smads during transforming growth factor β-induced epithelial-mesenchymal transition. Mol Med Rep 11 (3):2241–2247
241. Ogiso M, Shogomori H, Hoshi M (1998) Localization of LewisX, sialyl-LewisX and α-galactosyl epitopes on glycosphingolipids in lens tissues. Glycobiology 8:95–105
242. Ariga T, Tao RV, Lee BC, Yamawaki M, Yoshino H, Scarsdale NJ, Kasama T, Kushi Y, Yu RK (1994) Glycolipid composition of human cataractous lenses. Characterization of Lewisx glycolipids. J Biol Chem 269:2667–2675
243. TuA W, Luo K (2007) Acetylation of Smad2 by the co-activator p300 regulates activin and transforming growth factor β response. J Biol Chem 282:21187–21196
244. Datta PK, Blake MC, Moses HL (2000) Regulation of plasminogen activator inhibitor-1 expression by transforming growth factor-β-induced physical and functional interactions between smads and Sp1. J Biol Chem 275:40014–40019
245. Perdue N, Yan Q (2006) Caveolin-1 is up-regulated in transdifferentiated lens epithelial cells but minimal in normal human and murine lenses. Exp Eye Res 83:1154–1161
246. ThieryJ P, SleemanJ P (2006) Complex networks orchestrate epithelial–mesenchymal transitions. Nat Rev Mol Cell Biol 7:131–142
247. Zuo W, ChenYG (2009) Specific activation of mitogen-activated protein kinase by transforming growth factor-β receptors in lipid rafts is required for epithelial cell plasticity. Mol Biol Cell 20:1020–1029
248. Yao K, Tan J, Ye P, Wang K, Xu W, Shen TX, Tang X (2007) Integrin β1-mediated signaling is involved in transforming growth factor-β2-promoted migration in human lens epithelial cells. Mol Vis 13:1769–1776
249. Roberts AB, Tian F, Byfield SD, Stuelten C, Ooshima A, Saika S, Flanders KC (2006) Smad3 is key to TGF-β-mediated epithelial-to-mesenchymal transition, fibrosis, tumor suppression and metastasis. Cytokine Growth Factor Rev 17:19–27
250. Schmierer B, HillC S (2007) TGFβ-SMAD signal transduction: molecular specificity and functional flexibility. Nat Rev Mol Cell Biol 8:970–982

251. Tagami S, Inokuchi JI, Kabayama K, Yoshimura H, Kitamura F, Uemura S, Ogawa C, Ishii A, Saito M, Ohtsuka Y et al (2002) Ganglioside GM3 participates in the pathological conditions of insulin resistance. J Biol Chem 277:3085–3092
252. Kabayama K, Sato T, Saito K, Loberto N, Prinetti A, Sonnino S, Kinjo M, Igarashi Y, Inokuchi J (2007) Dissociation of the insulin receptor and caveolin-1 complex by ganglioside GM3 in the state of insulin resistance. Proc Natl Acad Sci U S A 104:13678–13683
253. Wang XQ, Yan Q, Sun P, Liu JW, Go L, McDaniel SM, Paller AS (2007) Suppression of epidermal growth factor receptor signaling by protein kinase C-α activation requires CD82, caveolin-1, and ganglioside. Cancer Res 67:9986–9995
254. Langeveld M, Aerts JM (2009) Glycosphingolipids and insulin resistance. Prog Lipid Res 48 (3–4):196–205
255. Sasaki A, Hata K, Suzuki S, Sawada M, Wada T, Yamaguchi K, Obinata M, Tateno H, Suzuki H, Miyagi T (2003) Overexpression of plasma membrane-associated sialidase attenuates insulin signaling in transgenic mice. J Biol Chem 278(30):27896–27902
256. Lipina C, Hundal HS (2015) Ganglioside GM3 as a gatekeeper of obesity-associated insulin resistance: evidence and mechanisms. FEBS Lett 589(21):3221–3227
257. Bremer EG, Hakomori S (1984) Gangliosides as receptor modulators. Adv Exp Med Biol 174:381–394
258. Sasaki N, Itakura Y, Toyoda M (2017) Ganglioside GM1 contributes to extracellular/intracellular regulation of insulin resistance, impairment of insulin signaling and down-stream eNOS activation, in human aortic endothelial cells after short- or long-term exposure to TNFα. Oncotarget 9(5):5562–5577
259. Vainio S, Heino S, Mansson JE, Fredman P, Kuismanen E, Vaarala O, Ikonen E (2002) Dynamic association of human insulin receptor with lipid rafts in cells lacking caveolae. EMBO Rep 3:95–100
260. Julien S, Bobowski M, Steenackers A, Le Bourhis X, Delannoy P (2003) Do gangliosides regulate RTKs signaling? Cell 2(4):751–767
261. Nishio M, Fukumoto S, Furukawa K, Ichimura A, Miyazaki H, Kusunoki S, Urano T, Furukawa K (2004) Overexpressed GM1 suppresses nerve growth factor (NGF) signals by modulating the intracellular localization of NGF receptors and membrane fluidity in PC12 cells. J Biol Chem 279:33368–33378
262. Malorni W, Giammarioli A, Garofalo T, Sorice M (2007) Dynamics of lipid raft components during lymphocyte apoptosis: the paradigmatic role of GD3. Apoptosis 12:941–949
263. Garofalo T, Misasi R, Mattei V, Giammarioli AM, Malorni W, Pontieri GM, Pavan A, Sorice M (2003) Association of the death-inducing signaling complex with microdomains after triggering through CD95/Fas. Evidence for caspase-8-ganglioside interaction in T cells. J Biol Chem 278:8309–8315
264. Legler DF, Micheau O, Doucey MA, Tschopp J, Bron C (2003) Recruitment of TNF receptor 1 to lipid rafts is essential for TNF alpha-mediated NF-kappaB activation. Immunity 18:655–664
265. Zhou J, Cox NR, Ewald SJ, Morrison NE, Baker HJ (1998) Evaluation of GM1 ganglioside-mediated apoptosis in feline thymocytes. Vet Immunol Immunopathol 66:25–42
266. Watanabe R, Ohyama C, Aoki H, Takahashi T, Satoh M, Saito S, Hoshi S, Ishii A, Saito M, Arai Y (2002) Ganglioside g(m3) overexpression induces apoptosis and reduces malignant potential in murine bladder cancer. Cancer Res 62:3850–3854
267. DeMaria R, Lenti L, Malisan F, d'Agostino F, Tomassini B, Zeuner A, Rippo MR, Testi R (1997) Requirement for GD3 ganglioside in CD95- and ceramide-induced apoptosis. Science 277:1652–1655
268. DeMaria R, Rippo MR, Schuchman EH, Testi R (1998) Acidic sphingomyelinase (ASM) is necessary for fas-induced GD3 ganglioside accumulation and efficient apoptosis of lymphoid cells. J Exp Med 187:897–902

269. Rippo MR, Malisan F, Ravagnan L, Tomassini B, Condo I, Costantini P, Susin SA, Rufini A, Todaro M, Kroemer G, Testi R (2000) GD3 ganglioside directly targets mitochondria in a bcl-2-controlled fashion. FASEB J 14:2047–2054
270. Colell A, Garcia-Ruiz C, Roman J, Ballesta A, Fernandez-Checa JC (2001) Ganglioside GD3 enhances apoptosis by suppressing the nuclear factor-kappa B-dependent survival pathway. FASEB J 15:1068–1070
271. Kang NY, Kang SK, Lee YC, Choi HJ, Lee YS, Cho SY, Kim YS, Ko JH, Kim CH (2006) Transcriptional regulation of the human GD3 synthase gene expression in Fas-induced Jurkat T cells: a critical role of transcription factor NF-kappaB in regulated expression. Glycobiology 16(5):375–389
272. Sorice M, Garofalo T, Misasi R, Manganelli V, Vona R, Malorni W (2012) Ganglioside GD3 as a raft component in cell death regulation. Anti Cancer Agents Med Chem 12(4):376–382
273. Peter ME, Budd RC, Desbarats J, Hedrick SM, Hueber AO, Newell MK, Owen LB, Pope RM, Tschopp J, Wajant H, Wallach D, Wiltrout RH, Zörnig M, Lynch DH (2007) The CD95 receptor: apoptosis revisited. Cell 129:447–450
274. Jacewicz M, Clausen H, Nudelman E, Donohue-Rolfe A, Keusch GT (1986) Pathogenesis of shigella diarrhea. XI. Isolation of a shigella toxin-binding glycolipid from rabbit jejunum and HeLa cells and its identification as globotriaosylceramide. J Exp Med 163:1391–1404
275. Lingwood CA, Law H, Richardson S, Petric M, Brunton JL, De Grandis S, Karmali M. 1987. J Biol Chem 262, 8834–8839
276. Gordon WR, Vardar-Ulu D, Histen G, Sanchez-Irizarry C, Aster JC, Blacklow SC (2007) Structural basis for autoinhibition of notch. Nat Struct Mol Biol 14:295–300
277. Andersson ER, Sandberg R, Lendahl U (2011) Notch signaling: simplicity in design, versatility in function. Development 138:3593–3612
278. Brou C (2009) Intracellular trafficking of notch receptors and ligands. Exp Cell Res 315 (9):1549–1555
279. Hamel S, Fantini J, Schweisguth F (2010) Notch ligand activity is modulated by glycosphingolipid membrane composition in Drosophila melanogaster. J Cell Biol 188:581–594
280. Sanchez-Niño MD, Carpio D, Sanz AB, Ruiz-Ortega M, Mezzano S, Ortiz A (2015) Lyso-Gb3 activates Notch1 in human podocytes. Hum Mol Genet 24(20):5720–5732
281. Merrill AH, Stokes TH, Momin A, Park H, Portz BJ, Kelly S, Wang E, Sullards MC, Wang MD (2009) Sphingolipidomics: a valuable tool for understanding the roles of sphingolipids in biology and disease. J Lipid Res 50(Suppl):S97–S102
282. Hannun YA, Obeid LM (2011) Many ceramides. J Biol Chem 286(32):27855–27862
283. Pierce KL, Premont RT, Lefkowitz RJ (2002) Seven-transmembrane receptors. Nat Rev Mol Cell Biol 3:639–650
284. Jacobson KA (2015) New paradigms in GPCR drug discovery. Biochem Pharmacol 98:541–555
285. Thomsen W, Frazer J, Unett D (2005) Functional assays for screening GPCR targets. Curr Opin Biotechnol 16:655–665
286. Jafurulla M, Chattopadhyay A (2015) Sphingolipids in the function of G protein-coupled receptors. Eur J Pharmacol 763:241–246
287. Singh P, Paila YD, Chattopadhyay A (2012) Role of glycosphingolipids in the function of human serotonin$_{1A}$ receptors. J Neurochem 123:716–724
288. Fantini J, Barrantes FJ (2009) Sphingolipid/cholesterol regulation of neurotransmitter receptor conformation and function. Biochim Biophys Acta 1788:2345–2361
289. Jin UH, Ha KT, Kim KW, Chang YC, Lee YC, Ko JH, Kim CH (2008) Membrane type sialidase inhibits the megakaryocytic differentiation of human leukemia K562 cells. Biochim Biophys Acta 1780(5):757–763
290. Shen K-F, Crain SM (1990) Cholera toxin-B subunit blocks opioid excitatory effects on sensory neuron action potentials indicating the GM1 ganglioside may regulate Gs-linked opioid receptor functions. Brain Res 531:1–7

291. Wu G et al (1997) Interaction of δ-opioid receptor with GM1 ganglioside: conversion from inhibitory to excitatory mode. Mol Brain Res 44:341–346
292. Wu G et al (1998) The role of GM1 ganglioside in regulating excitatory opioid effects. Ann N Y Acad Sci 845:126–138
293. Saito M et al (1995) Modulation by GM1 ganglioside of 1-adrenergic receptor-induced cyclic AMP formation in Sf9 cells. Biochem Biophys Acta 1267:1–5
294. Wu G et al (1996) GM1 ganglioside modulates prostaglandin E1 stimulated adenylyl cyclase in Neuro-2A cells. Glycoconj J 13:235–239
295. Chattopadhyay A, Paila YD, Shrivastava S, Tiwari S, Singh P, Fantini J (2012) Sphingolipid-binding domain in the serotonin$_{1A}$ receptor. Adv Exp Med Biol 749:279–293
296. Fantini J, Yahi N (2015) Lipid regulation of receptor function. In: Fantini J, Yahi N (eds) Brain lipids in synaptic function and neurological disease. Elsevier, San Diego, pp 163–181
297. Prasanna X, Jafurulla M, Sengupta D, Chattopadhyay A (2016) The ganglioside GM1 interacts with the serotonin1A receptor via the sphingolipid binding domain. Biochim Biophys Acta 1858(11):2818–2826
298. Imai Y, Soda M, Inoue H, Hattori N, Mizuno Y, Takahashi R (2001) An unfolded putative transmembrane polypeptide, which can lead to endoplasmic reticulum stress, is a substrate of parkin. Cell 105:891–902
299. Lundius EG, Vukojevic V, Hertz E, Stroth N, Cederlund A, Hiraiwa M, Terenius L, Svenningsson P (2014) GPR37 protein trafficking to the plasma membrane regulated by prosaposin and GM1 gangliosides promotes cell viability. J Biol Chem 289(8):4660–4673
300. Sjögren B, Svenningsson P (2007) Depletion of the lipid raft constituents, sphingomyelin and ganglioside, decreases serotonin binding at human 5-HT7a receptors in HeLa cells. Acta Physiol 190:47–53
301. Prasanna X, Jafurulla M, Sengupta D, Chattopadhyay A (2016) The ganglioside GM1 interacts with the serotonin 1A receptor via the sphingolipid binding domain. Biochim Biophys Acta 1858(11):2818–2826
302. Ha SH, Kang SK, Choi H, Kwak CH, Abekura F, Park JY, Kwon KM, Chang HW, Lee YC, Ha KT, Hou BK, Chung TW, Kim CH (2017) Induction of GD3/α1-adrenergic receptor/transglutaminase 2-mediated erythroid differentiation in chronic myelogenous leukemic K562 cells. Oncotarget 8(42):72205–72219
303. Zhang J, Lesort M, Guttmann RP, Johnson GV (1998) Modulation of the in situ activity of tissue transglutaminase by calcium and GTP. J Biol Chem 273:2288–2295
304. Kang SK, Lee JY, Chung TW, Kim CH (2004) Overexpression of transglutaminase 2 accelerates the erythroid differentiation of human chronic myelogenous leukemia K562 cell line through PI3K/Akt signaling pathway. FEBS Lett 577:361–366
305. Kang NY, Kim CH, Kim KS, Ko JH, Lee JH, Jeong YK, Lee YC (2007) Expression of the human CMP-NeuAc: GM3 alpha2,8-sialyltransferase (GD3 synthase) gene through the NF-kappaB activation in human melanoma SK-MEL-2 cells. Biochim Biophys Acta 1769:622–630
306. Ha KT, Lee YC, Kim CH (2004) Overexpression of GD3 synthase induces apoptosis of vascular endothelial ECV304 cells through downregulation of Bcl-2. FEBS Lett 568:183–187
307. Steinert S, Lee E, Tresset G, Zhang D, Hortsch R, Wetzel R, Hebbar S, Sundram JR, Kesavapany S, Boschke E, Kraut R (2008) A fluorescent glycolipid-binding peptide probe traces cholesterol dependent microdomain-derived trafficking pathways. PLoS One 3:e2933
308. Fantini J, Yahi N, Garmy N (2013) Cholesterol accelerates the binding of Alzheimer's β-amyloid peptide to ganglioside GM1 through a universal hydrogen-bond-dependent sterol tuning of glycolipid conformation. Front Physiol 2013:4
309. Wang Y, Kraut R, Mu Y (2015) Aβ1-25-derived Sphingolipid-domain tracer peptide SBD interacts with membrane Ganglioside clusters via a coil-helix-coil motif. Int J Mol Sci 16 (11):26318–26332
310. Boccuto L, Aoki K, Flanagan-Steet H, Chen CF, Fan X, Bartel F, Petukh M, Pittman A, Saul R, Chaubey A, Alexov E, Tiemeyer M, Steet R, Schwartz CE (2014) A mutation in a

ganglioside biosynthetic enzyme, ST3GAL5, results in salt and pepper syndrome, a neurocutaneous disorder with altered glycolipid and glycoprotein glycosylation. Hum Mol Genet 23:418–433
311. Hanley JG (2014) Subunit-specific trafficking mechanisms regulating the synaptic expression of Ca-permeable AMPA receptors. Semin Cell Dev Biol 27:14–22
312. Zhang J, Wang Y, Chi Z, Keuss MJ, Pai YM, Kang HC, Shin JH, Bugayenko A, Wang H, Xiong Y, Pletnikov MV, Mattson MP, Dawson TM, Dawson VL (2011) The AAA+ ATPase Thorase regulates AMPA receptor-dependent synaptic plasticity and behavior. Cell 145:284–299
313. Prendergast J, Umanah GK, Yoo SW, Lagerlof O, Motari MG, Cole RN, Huganir RL, Dawson TM, Dawson VL, Schnaar RL (2014) Ganglioside regulation of AMPA receptor trafficking. J Neurosci 34(39):13246–13258
314. Cole AA, Dosemeci A, Reese TS (2010) Co-segregation of AMPA receptors with GM1 ganglioside in synaptosomal membrane subfractions. Biochem J 427:535–540
315. Rogawski MA (2913) AMPA receptors as a molecular target in epilepsy therapy. Acta Neurol Scand Suppl 197:9–18
316. Anggono V, Huganir RL (2012) Regulation of AMPA receptor trafficking and synaptic plasticity. Curr Opin Neurobiol 22:461–469
317. Bomsel M, Alfsen A (2003) Review entry of viruses through the epithelial barrier: pathogenic trickery. Nat Rev Mol Cell Biol 4(1):57–68
318. Olofsson S, Bergstrom T (2005) Glycoconjugate glycans as viral receptors. Ann Med 37:154–172
319. Greber UF (2002) Signaling in viral entry. Cell Mol Life Sci 59:608–626
320. Neu U, Stehle T, Atwood WJ (2009) The Polyomaviridae: contributions of virus structure to our understanding of virus receptors and infectious entry. Virology 384:389–399
321. Taube S, Jiang M, Wobus CE (2010) Glycosphingolipids as receptors for non-enveloped viruses. Viruses 2:1011–1049
322. Villar E, Barroso IM (2006) Role of sialic acid-containing molecules in paramyxovirus entry into the host cell: a minireview. Glycoconj J 23:5–17
323. Choppin PW, Compans RW (1975) Reproduction of paramyxoviruses. In: Fraenkel-Conrat H, Wagner RR (eds) Comprehensive virology. Plenum Press, New York, pp 95–178
324. Lamb RA, Kolakofsky D (2001) Paramyxoviridae: the viruses and their replication. In: Fields BN, Knippe DM, Kato A (eds) Fundamental virology. Lippincott-Raven, New York, pp 1305–1340
325. Schmidt M, Chiorini JA (2006) Gangliosides are essential for bovine adeno-associated virus entry. J Virol 80:5516–5522
326. Bergelson LD, Bukrinskaya AG, Prokazova NV, Shaposhnikova GI, Kocharov SL, Shevchenko VP, Kornilaeva GV, Fomina-Ageeva EV (1982) Role of gangliosides in reception of influenza virus. Eur J Biochem 128:467–474
327. Taube S, Perry JW, Yetming K, Patel SP, Auble H, Shu L, Nawar HF, Lee CH, Connell TD, Shayman JA, Wobus CE (2009) Ganglioside-linked terminal sialic acid moieties on murine macrophages function as attachment receptors for murine noroviruses. J Virol 83:4092–4101
328. Rolsma MD, Kuhlenschmidt TB, Gelberg HB, Kuhlenschmidt MS (1998) Structure and function of a ganglioside receptor for porcine rotavirus. J Virol 72:9079–9091
329. Guo CT, Nakagomi O, Mochizuki M, Ishida H, Kiso M, Ohta Y, Suzuki T, Miyamoto D, Hidari KI, Suzuki Y (1999) Ganglioside GM(1a) on the cell surface is involved in the infection by human rotavirus KUN and MO strains. J Biochem 126:683–688
330. Schnaar RL (1991) Glycosphingolipids in cell surface recognition. Glycobiology 1:477–485
331. Suzuki T, Portner A, Scroggs RA, Uchikawa M, Koyama N, Matsuo K, Suzuki Y, Takimoto T (2001) Receptor specificities of human respiroviruses. J Virol 75:4604–4613
332. Suzuki Y, Suzuki T, Matsunaga M, Matsumoto M (1985) Gangliosides as paramyxovirus receptor. Structural requirement of sialo-oligosaccharides in receptors for hemagglutinating virus of Japan (Sendai virus) and Newcastle disease virus. J Biochem 97:1189–1199

333. Wybenga LE, Epand RF, Nir S, Chu JW, Sharom FJ, Flanagan TD, Epand RM (1996) Glycophorin as a receptor for Sendai virus. Biochemistry 35(29):9513–9518
334. Markwell MA, Svennerholm L, Paulson JC (1981) Specific gangliosides function as host cell receptors for Sendai virus. Proc Natl Acad Sci U S A 78:5406–5410
335. Ferreira L, Villar E, Muñoz-Barroso I (2004) Gangliosides and N-glycoproteins function as Newcastle disease virus receptors. Int J Biochem Cell Biol 36(11):2344–2356
336. Lafon M (2005) Rabies virus receptors. J Neurovirol 11(1):82–87
337. Superti F, Hauttecoeur B, Morelec MJ, Goldoni P, Bizzini B, Tsiang H (1986) Involvement of gangliosides in rabies virus infection. J Gen Virol 67:47–56
338. Hrzenjak T, Roguljić A, Kasuba V, Karlović M (1994) The role of sialic acid and Ca2+ ions in rabies virus-infected foxes brain. Comp Immunol Microbiol Infect Dis 17(1):71–75
339. Rogers GN, Paulson JC, Daniels RS, Skehel JJ, Wilson IA, Wiley DC (1983) Single amino acid substitutions in influenza haemagglutinin change receptor binding specificity. Nature 304:76–78
340. Haywood AM (1994) Virus receptors: binding, adhesion strengthening. and changes in viral structure J Virol 68:1–5
341. Barton ES, Connolly JL, Forrest JC, Chappell JD, Dermody TS (2001) Utilization of sialic acid as a coreceptor enhances reovirus attachment by multistep adhesion strengthening. J Biol Chem 276:2200–2211
342. Coffin J, Haase A, Levy JA, Montagnier L, Oroszlan S, Teich N, Temin H, Toyoshima K, Varmus H, Vogt P (1986) What to call the AIDS virus? Nature 321(6065):10
343. Gallo SA, Finnegan CM, Viard M, Raviv Y, Dimitrov A, Rawat SS, Puri A, Durell S, Blumenthal R (2003) The HIV Env-mediated fusion reaction. Biochim Biophys Acta 1614 (1):36–50
344. Bhat S, Spitalnik SL, Gonzalez-Scarano F, Silberberg DH (1991) Galactosyl ceramide or a derivative is an essential component of the neural receptor for human immunodeficiency virus type 1 envelope glycoprotein gp120. Proc Natl Acad Sci U S A 88(16):7131–7134
345. Yahi N, Baghdiguian S, Moreau H, Fantini J (1992) Galactosyl ceramide (or a closely related molecule) is the receptor for human immunodeficiency virus type 1 on human colon epithelial HT29 cells. J Virol 66(8):4848–4854
346. Erikson E, Wratil PR, Frank M, Ambiel I, Pahnke K, Pino M et al (2015) Mouse Siglec-1 mediates trans-infection of surface-bound murine Leukemia virus in a Sialic acid N-acyl side chain-dependent manner. J Biol Chem 290(45):27345–27359
347. Hammonds JE, Beeman N, Ding L, Takushi S, Francis AC, Wang JJ, Melikyan GB, Spearman P (2017) Siglec-1 initiates formation of the virus-containing compartment and enhances macrophage-to-T cell transmission of HIV-1. PLoS Pathog 13(1):e1006181
348. Berger EA, Murphy PM, Farber JM (1999) Chemokine receptors as HIV-1 coreceptors: roles in viral entry, tropism, and disease. Annu Rev Immunol 17:657–700
349. Nehete PN, Vela EM, Hossain MM, Sarkar AK, Yahi N, Fantini J, Sastry KJ (2002) A post-CD4-binding step involving interaction of the V3 region of viral gp120 with host cell surface glycosphingolipids is common to entry and infection by diverse HIV-1 strains. Antivir Res 56 (3):233–251
350. Lund N, Olsson ML, Ramkumar S, Sakac D, Yahalom V, Levene C, Hellberg A, Ma XZ, Binnington B, Jung D, Lingwood CA, Branch DR (2009) The human P(k) histo-blood group antigen provides protection against HIV-1 infection. Blood 113:4980–4991
351. Branch DR (2010) Blood groups and susceptibility to virus infection: new developments. Curr Opin Hematol 17(6):558–564
352. Lingwood CA, Branch DR (2011) The role of glycosphingolipids in HIV/AIDS. Discov Med 11(59):303–313
353. Lingwood CA (2003) Shiga toxin receptor glycolipid binding: pathology and utility. In: Philpott D, Ebel F (eds) Methods in molecular medicine. Shiga toxin-producing Escherichia coli: methods and protocols. Humana Press, Totowa, NJ, pp 165–186

354. Bachan S, Fantini J, Joshi A, Garg H, Mootoo DR (2011) Synthesis, gp120 binding and anti-HIV activity of fatty acid esters of 1,1-linked disaccharides. Bioorg Med Chem 19 (16):4803–4811
355. Delezay O, Hammache D, Fantini J, Yahi N (1996) SPC3, a V3 loop-derived synthetic peptide inhibitor of HIV-1 infection, binds to cell surface glycosphingolipids. Biochemistry 35 (49):15663–15671
356. Fantini J, Yahi N (2011) Molecular basis for the glycosphingolipid-binding specificity of alpha-synuclein: key role of tyrosine 39 in membrane insertion. J Mol Biol 408(4):654–669
357. Chattopadhyay A, Paila YD, Shrivastava S, Tiwari S, Singh P, Fantini J (2012) Sphingolipid-binding domain in the serotonin(1A) receptor. Adv Exp Med Biol 749:279–293
358. Maresca M, Derghal A, Caravagna C, Dudin S, Fantini J (2008) Controlled aggregation of adenine by sugars: physicochemical studies, molecular modelling simulations of sugar-aromatic CH-pi stacking interactions, and biological significance. Phys Chem Chem Phys 10(19):2792–2800
359. Neu U, Woellner K, Gauglitz G, Stehle T (2008) Structural basis of GM1 ganglioside recognition by simian virus 40. Proc Natl Acad Sci U S A 105(13):5219–5224
360. Geoghegan EM, Pastrana DV, Schowalter RM, Ray U, Gao W, Ho M, Pauly GT, Sigano DM, Kaynor C (2017) Cahir-McFarland E, Combaluzier B, Grimm J, Buck CB. 2017. Infectious entry and neutralization of pathogenic JC Polyomaviruses. Cell Rep 21(5):1169–1179
361. Magaldi TG, Buch MH, Murata H, Erickson KD, Neu U, Garcea RL, Peden K, Stehle T, DiMaio D (2012) Mutations in the GM1 binding site of simian virus 40 VP1 alter receptor usage and cell tropism. J Virol 86(13):7028–7042
362. Campanero-Rhodes MA, Smith A, Chai W, Sonnino S, Mauri L, Childs RA, Zhang Y, Ewers H, Helenius A, Imberty A, Feizi T (2007) N-glycolyl GM1 ganglioside as a receptor for simian virus 40. J Virol 81:12846–12858
363. Neu U, Maginnis MS, Palma AS, Ströh LJ, Nelson CD, Feizi T, Atwood WJ, Stehle T (2010) Structure-function analysis of the human JC polyomavirus establishes the LSTc pentasaccharide as a functional receptor motif. Cell Host Microbe 8:309–319
364. Ströh LJ, Maginnis MS, Blaum BS, Nelson CD, Neu U, Gee GV, O'Hara BA, Motamedi N, DiMaio D, Atwood WJ, Stehle T (2015) The greater affinity of JC Polyomavirus capsid for alpha2,6-linked Lactoseries Tetrasaccharide c than for other Sialylated Glycans is a major determinant of infectivity. J Virol 89:6364–6375
365. Dugan AS, Gasparovic ML, Atwood WJ (2008) Direct correlation between sialic acid binding and infection of cells by two human polyomaviruses (JC virus and BK virus). J Virol 82 (5):2560–2564
366. O'Hara SD, Garcea RL (2016) Murine Polyomavirus cell surface receptors activate distinct Signaling pathways required for infection. MBio 7(6):e01836–e01816
367. Dugan AS, Eash S, Atwood WJ (2005) An N-linked glycoprotein with alpha(2,3)-linked sialic acid is a receptor for BK virus. J Virol 79(22):14442–14445
368. Neu U, Hengel H, Blaum BS, Schowalter RM, Macejak D, Gilbert M, Wakarchuk WW, Imamura A, Ando H, Kiso M, Arnberg N, Garcea RL, Peters T, Buck CB, Stehle T (2012) Structures of Merkel cell polyomavirus VP1 complexes define a sialic acid binding site required for infection. PLoS Pathog 8:e1002738
369. Neu U, Khan ZM, Schuch B, Palma AS, Liu Y, Pawlita M, Feizi T, Stehle T (2013) Structures of B-lymphotropic polyomavirus VP1 in complex with oligosaccharide ligands. PLoS Pathog 9:e1003714
370. Khan ZM, Liu Y, Neu U, Gilbert M, Ehlers B, Feizi T, Stehle T (2014) Crystallographic and glycan microarray analysis of human Polyomavirus 9 VP1 identifies N-glycolyl neuraminic acid as a receptor candidate. J Virol 88(11):6100–6111
371. Ströh LJ, Neu U, Blaum BS, Buch MHC, Garcea RL, Stehle T (2014) Structure analysis of the major capsid proteins of human polyomaviruses 6 and 7 reveals an obstructed sialic acid binding site. J Virol 88:10831–10839

372. You J, O'Hara SD, Velupillai P, Castle S, Levery S, Garcea RL, Benjamin T (2015) Ganglioside and non-ganglioside mediated host responses to the mouse polyomavirus. PLoS Pathog 11:e1005175
373. Qian M, Cai D, Verhey KJ, Tsai B (2009) A lipid receptor sorts polyomavirus from the endolysosome to the endoplasmic reticulum to cause infection. PLoS Pathog 5:e1000465
374. van der Meijden E, Janssens RW, Lauber C, Bouwes Bavinck JN, Gorbalenya AE, Feltkamp MC (2010) Discovery of a new human polyomavirus associated with trichodysplasia spinulosa in an immunocompromized patient. PLoS Pathog 6(7):e1001024
375. Ströh LJ, Gee GV, Blaum BS, Dugan AS, Feltkamp MC, Atwood WJ, Stehle T (2015) Trichodysplasia spinulosa-associated polyomavirus uses a displaced binding site on VP1 to engage sialylated glycolipids. PLoS Pathog 11(8):e1005112
376. Anastasia L, Holguera J, Bianchi A, D'Avila F, Papini N, Tringali C, Monti E, Villar E, Venerando B, Muñoz-Barroso I, Tettamanti G (2008) Over-expression of mammalian sialidase NEU3 reduces Newcastle disease virus entry and propagation in COS7 cells. Biochim Biophys Acta 1780(3):504–512
377. Nilsson EC, Storm RJ, Bauer J, Johansson SM, Lookene A, Ångström J, Hedenström M, Eriksson TL, Frängsmyr L, Rinaldi S, Willison HJ, Pedrosa Domellöf F, Stehle T, Arnberg N (2011) The GD1a glycan is a cellular receptor for adenoviruses causing epidemic keratoconjunctivitis. Nat Med 17(1):105–109
378. Arnberg N, Edlund K, Kidd AH, Wadell G (2000) Adenovirus type 37 uses sialic acid as a cellular receptor. J Virol 74(1):42–48
379. Santos AX, Maia JE, Crespo PM, Pettenuzzo LF, Daniotti JL, Barbé-Tuana FM, Martins LM, Trindade VM, Borojevic R, Guma FC (2011) GD1a modulates GM-CSF-induced cell proliferation. Cytokine 56(3):600–607
380. von Itzstein M, Wu WY, Kok GB, Pegg MS, Dyason JC, Jin B, Van Phan T, Smythe ML, White HF, Oliver SW (1993) Rational design of potent sialidase-based inhibitors of influenza virus replication. Nature 363:418–423
381. Kim CU, Lew W, Williams MA, Liu H, Zhang L, Swaminathan S, Bischofberger N, Chen MS, Mendel DB, Tai CY, Laver WG, Stevens RC (1997) Influenza neuraminidase inhibitors possessing a novel hydrophobic interaction in the enzyme active site: design, synthesis, and structural analysis of carbocyclic sialic acid analogues with potent anti-influenza activity. J Am Chem Soc 119:681–690
382. Di Pasquale G, Kaludov N, Agbandje-McKenna M, Chiorini JA (2010) BAAV transcytosis requires an interaction with beta-1-4 linked- glucosamine and gp96. PLoS One 5(3):e9336
383. Harbison CE, Chiorini JA, Parrish CR (2008) The parvovirus capsid odyssey: from the cell surface to the nucleus. Trends Microbiol 16(5):208–214
384. Coulson BS (2015) Expanding diversity of glycan receptor usage by rotaviruses. Curr Opin Virol 15:90–96
385. Haselhorst T, Fiebig T, Dyason JC, Fleming FE, Blanchard H, Coulson BS, von Itzstein M (2011) Recognition of the GM3 ganglioside glycan by rhesus rotavirus particles. Angew Chem Int Ed Engl 50:1055–1058
386. Rodriguez JM, Chichon FJ, Martin-Forero E, Gonzalez-Camacho F, Carrascosa JL, Caston JR, Luque D (2014) New insights into rotavirus entry machinery: stabilization of rotavirus spike conformation is independent of trypsin cleavage. PLoS Pathog 10:e1004157
387. Civra A, Giuffrida MG, Donalisio M, Napolitano L, Takada Y, Coulson BS, Conti A, Lembo D (2015) Identification of equine lactadherin-derived peptides that inhibit rotavirus infection via integrin receptor competition. J Biol Chem 290:12403–12414
388. Graham KL, Fleming FE, Halasz P, Hewish MJ, Nagesha HS, Holmes IH, Takada Y, Coulson BS (2005) Rotaviruses interact with a4b7 and a4b1 integrins by binding the same integrin domains as natural ligands. J Gen Virol 86:3397–3408
389. Marionneau S, Cailleau-Thomas A, Rocher J, Le Moullac-Vaidye B, Ruvoen N, Clement M, Le Pendu J (2001) ABH and Lewis histo-blood group antigens, a model for the meaning of oligosaccharide diversity in the face of a changing world. Biochimie 83:565–573

390. Martinez MA, Lopez S, Arias CF, Isa P (2012) Gangliosides have a functional role during rotavirus cell entry. J Virol 87:1115–1122
391. Huang P, Xia M, Tan M, Zhong W, Wei C, Wang L, Morrow A, Jiang X (2012) Spike protein VP8* of human rotavirus recognizes histo-blood group antigens in a type-specific manner. J Virol 86:4833–4843
392. Fleming FE, Bohm R, Dang VT, Holloway G, Haselhorst T, Madge PD, Deveryshetty J, Yu X, Blanchard H, von Itzstein M (2014) Relative roles of GM1 ganglioside, N-acylneuraminic acids, and a2b1 integrin in mediating rotavirus infection. J Virol 88:4558–4571
393. Yu X, Dang VT, Fleming FE, von Itzstein M, Coulson BS, Blanchard H (2012) Structural basis of rotavirus strain preference towards N-acetyl- or N-glycolylneuraminic acid-containing receptors. J Virol 86:13456–13466
394. Rolsma DM, Gelberg HB, Kuhlenschmidt MS (1994) Assay for evaluation of rotavirus-cell interactions: identification of an enterocyte ganglioside fraction that mediates group a porcine rotavirus recognition. J Virol 68:258–268
395. Reiter DM, Frierson JM, Halvorson EE, Kobayashi T, Dermody TS (2011) Crystal structure of reovirus attachment protein sigma1 in complex with sialylated oligosaccharides. PLoS Pathog 7:e1002166
396. Superti F, Donelli G (1991) Gangliosides as binding sites in SA-11 rotavirus infection of LLC-MK2 cells. J Gen Virol 72(Pt 10):2467–2474
397. Delorme C, Brüssow H, Sidoti J, Roche N, Karlsson KA, Neeser JR, Teneberg S (2001) Glycosphingolipid binding specificities of rotavirus: identification of a sialic acid-binding epitope. J Virol 75(5):2276–2287
398. Martínez MA, López S, Arias CF, Isa P (2013) Gangliosides have a functional role during rotavirus cell entry. J Virol 87(2):1115–1122
399. Reiss K, Stencel JE, Liu Y, Blaum BS, Reiter DM, Feizi T, Dermody TS, Stehle T (2012) The GM2 glycan serves as a functional coreceptor for serotype 1 reovirus. PLoS Pathog 8(12): e1003078
400. Danthi P, Holm GH, Stehle T, Dermody TS (2013) Reovirus receptors, cell entry, and proapoptotic signaling. Adv Exp Med Biol 790:42–71
401. Adair RA, Roulstone V, Scott KJ, Morgan R, Nuovo GJ (2012) Cell carriage, delivery, and selective replication of an oncolytic virus in tumor in patients. Sci Transl Med 4:138ra177
402. Weiner HL, Powers ML, Fields BN (1980) Absolute linkage of virulence and central nervous system cell tropism of reoviruses to viral hemagglutinin. J Infect Dis 141:609–616
403. Stencel-Baerenwald J, Reiss K, Blaum BS, Colvin D, Li XN, Abel T, Boyd K, Stehle T, Dermody TS (2015) Glycan engagement dictates hydrocephalus induction by serotype 1 reovirus. MBio 6(2):e02356
404. Gomatos PJ, Tamm I (1962) Reactive sites of reovirus type 3 and their interaction with receptor substances. Virology 17:455–461
405. Turnbull WB, Precious BL, Homans SW (2004) Dissecting the cholera toxin-ganglioside GM1 interaction by isothermal titration calorimetry. J Am Chem Soc 126:1047–1054
406. Chen R, Jiang X, Sun D, Han G, Wang F, Ye M, Wang L, Zou H (2009) Glycoproteomics analysis of human liver tissue by combination of multiple enzyme digestion and hydrazide chemistry. J Proteome Res 8:651–661
407. Guglielmi KM, Kirchner E, Holm GH, Stehle T, Dermody TS (2007) Reovirus binding determinants in junctional adhesion molecule-a. J Biol Chem 282:17930–17940
408. Williams RK, Straus SE (1997) Specificity and affinity of binding of herpes simplex virus type 2 glycoprotein B to glycosaminoglycans. J Virol 71:1375–1380
409. Doms RW, Peipert SC (1997) Unwelcomed guests with master keys: how HIV uses chemokine receptors for cellular entry. Virology 235:179–190
410. Rogo LD, Mokhtari-Azad T, Kabir MH, Rezaei F (2014) Human parvovirus B19: a review. Acta Virol 58(3):199–213

411. Anderson MJ, Lewi E, Kidd IM, Hall SM, Cohen BJ (1984) An outbreak of erythema infectiosum associated with human parvovirus infection. J Hyg 93:85–93
412. Weigel-Kelley KA, Yoder MC, Srivastava A (2001) Recombinant human parvovirus B19 vectors: erythrocyte P antigen is necessary but not sufficient for successful transduction of human hematopoietic cells. J Virol 75:4110–4116
413. Brown KE, Young NS, Alving BM, Barbosa LH (2001) Parvovirus B19 implications for transfusion medicine. Summary of a workshop. Transfusion 41:130–135
414. Munakata Y, Saito-Ito T, Kumura-Ishii K, Huang J, Kodera T, Ishii T, Hirabayashi Y, Koyanagi Y, Sasaki T (2005) Ku80 autoantigen as a cellular coreceptor for human parvovirus B19 infection. Blood 106:3449–3456
415. Bönsch C, Zuercher C, Lieby P, Kempf C, Ros C (2010) The globoside receptor triggers structural changes in the B19 virus capsid that facilitate virus internalization. J Virol 84:11737–11746
416. Leisi R, Ruprecht N, Kempf C, Ros C (2013) Parvovirus B19 uptake is a highly selective process controlled by VP1u, a novel determinant of viral tropism. J Virol 87:13161–13167
417. Nasir W, Nilsson J, Olofsson S, Bally M, Rydell GE (2014) Parvovirus B19 VLP recognizes globoside in supported lipid bilayers. Virology 456-457:364–369
418. Matrosovich M, Herrler G, Klenk HD (2015) Sialic acid receptors of viruses. Top Curr Chem 367:1–28
419. Ito T, Suzuki Y, Mitnaul L, Vines A, Kida H, Kawaoka Y (1997) Receptor specificity of influenza a viruses correlates with the agglutination of erythrocytes from different animal species. Virology 227:493–499
420. Matrosovich MN, Krauss S, Webster RG (2001) H9N2 influenza a virus from poultry in Asia have human virus-like receptor specificity. Virology 281:156–162
421. Cardona CJ, Xing Z, Sandrock CE, Davis CE (2009) Avian influenza in birds and mammals. Comp Immunol Microbiol Infect Dis 32:255–273
422. Hirst GK (1941) The agglutination of red cells by allantoic fluid of chick embryos infected with influenza virus. Science 94:22–23
423. McClelland L, Hare R (1941) The adsorption of influenza virus by red cells and a new in vitro method of measuring antibodies for influenza virus. Can J Public Health 32:530–538
424. Peiris JS, Hui KP, Yen HL (2010) Host response to influenza virus: protection versus immunopathology. Curr Opin Immunol 22(4):475–481
425. Song L (2014) It is unlikely that influenza viruses will cause a pandemic again like what happened in 1918 and 1919. Front Public Health 7(2):39. https://doi.org/10.3389/fpubh.2014.00039. eCollection 2014
426. Ibricevic A, Pekosz A, Walter MJ, Newby C, Battaile JT, Brown EG, Holtzman MJ, Brody SL (2006) Influenza virus receptor specificity and cell tropism in mouse and human airway epithelial cells. J Virol 80:7469–7480
427. Horimoto K, Kawaoka Y (2005) Influenza: lessons from past pandemics, warnings from current incidents. Nat Rev Microbiol 3:591–600
428. Neumann G, Noda T, Kawaoka Y (2009) Emergence and pandemic potential of swine-origin H1N1 influenza virus. Nature 459:931–939
429. Thompson CI, Barclay WS, Zambon MC, Pickles RJ (2006) Infection of human airway epithelium by human and avian strains of influenza a virus. J Virol 80:8060–8068
430. Xie Y, Gong J, Li M, Fang H, Xu W (2011) The medicinal potential of influenza virus surface proteins: hemagglutinin and neuraminidase. Curr Med Chem 18:1050–1066
431. Kuiken T, Holmes EC, McCauley J, Rimmelzwaan GF, Williams CS, Grenfell BT (2006) Host species barriers to influenza virus infections. Science 312:394397
432. Guo CT, Takahashi N, Yagi H, Kato K, Takahashi T, Yi SQ, Chen Y, Ito T, Otsuki K, Kida H, Kawaoka Y, Hidari KI, Miyamoto D, Suzuki T, Suzuki Y (2007) The quail and chicken intestine have sialyl-galactose sugar chains responsible for the binding of influenza a viruses to human type receptors. Glycobiology 17:713–724

433. Auewarakul P, Suptawiwat O, Kongchanagul A, Sangma C, Suzuki K, Ungchusak S, Louisirirotchanakul H, Lerdsamran P, Pooruk A, Thitithanyanont C, Pittayawonganon C, Guo T, Hiramatsu H, Jampangern W, Chunsutthiwat S, Puthavathana P (2007) An avian influenza H5N1 virus that binds to a human-type receptor. J Virol 81:9950–9905
434. Stevens J, Blixt O, Chen LM, Donis RO, Paulson JC, Wilson IA (2008) Recent avian H5N1 viruses exhibit increased propensity for acquiring human receptor specificity. J Mol Biol 381:1382–1394
435. Oshansky CM, Pickens JA, Bradley KC, Jones LP, Saavedra-Ebner GM, Barber JP, Crabtree JM, Steinhauer DA, Tompkins SM, Tripp RA (2011) Avian influenza viruses infect primary human bronchial epithelial cells unconstrained by sialic acid α2,3 residues. PLoS One 6 (e21):183
436. Son SW, Song KH, Kwon HY, Kim KS, Kim CH, Oh KB, Choo YK, Lee YC (2011) Transcriptional activation of pig Galβ1,3GalNAc α2,3-sialyltransferase (pST3Gal I) gene by TGF-β1 in porcine kidney PK-15 cells. Biochem Biophys Res Commun 414(1):159–164
437. Ko HK, Song KH, Jin UH, Seong HH, Chang YC, Kim NH, Kim DS, Lee YC, Kim CH (2010) Molecular characterization of pig alpha2,3-gal-beta1,3-GalNAc-alpha2,6-sialyltransferase (pST6GalNAc IV) gene specific for Neu5Acalpha2-3Galbeta1-3GalNAc trisaccharide structure. Glycoconj J 27(3):367–374
438. Vrijens P, Noppen S, Boogaerts T, Vanstreels E, Ronca R, Chiodelli P, Laporte M, Vanderlinden E, Liekens S, Stevaert A, Naesens L (2019) 2019. Influenza virus entry via the GM3 ganglioside-mediated platelet-derived growth factor receptor β signalling pathway. J Gen Virol 100(4):583–601
439. Baldauf KJ, Royal JM, Hamorsky KT, Matoba N (2015) Cholera toxin B: one subunit with many pharmaceutical applications. Toxins (Basel) 7(3):974–996
440. Fukushima K, Takahashi T, Ueyama H, Takaguchi M, Ito S, Oishi K, Minami A, Ishitsubo E, Tokiwa H, Takimoto T, Suzuki T (2015) Amino acid substitutions contributing to α2,6-sialic acid linkage binding specificity of human parainfluenza virus type 3 hemagglutinin-neuraminidase. FEBS Lett. S0014-5793(15)00238-0
441. Martinez M, Joffraud M, Giraud S, Baïsse B, Bernimoulin MP, Schapira M, Spertini O (2005) Regulation of PSGL-1 interactions with L-selectin, P-selectin, and E-selectin: role of human fucosyltransferase-IV and -VII. J BiolChem 280(7):5378–5390
442. Wu WL, Lau SY, Chen Y, Wang G, Mok BW, Wen X, Wang P, Song W, Lin T, Chan KH, Yuen KY, Chen H (2012) The 2008-2009 H1N1 influenza virus exhibits reduced susceptibility to antibody inhibition: implications for the prevalence of oseltamivir resistant variant viruses. Antivir Res 93:144–153
443. Xie Y, Huang B, Yu K, Shi F, Liu T, Xu W (2013) Caffeic acid derivatives: a new type of influenza neuraminidase inhibitors. Bioorg Med Chem Lett 23(12):3556–3560
444. Liu AL, Wang HD, Lee SM, Wang YT, Du GH (2008) Structure-activity relationship of flavonoids as influenza virus neuraminidase inhibitors and their in vitro anti-viral activities. Bioorg Med Chem 16:7141–7147
445. Ryu YB, Curtis-Long MJ, Lee JW, Kim JH, Kim JY, Kang KY, Lee WS, Park KH (2009) Characteristic of neuraminidase inhibitory xanthones from Cudrania tricuspidata. Bioorg Med Chem 17:2744–2750
446. Dao TT, Dang TT, Nguyen PH, Kim E, Thuong PT, Oh WK (2012) Xanthones from polygala karensium inhibit neuraminidases from influenza a viruses. Bioorg Med Chem Lett 22:3688–3692
447. Takano R, Kiso M, Igarashi M, Le QM, Sekijima M, Ito K, Takada A, Kawaoka Y (2013) Molecular mechanisms underlying oseltamivir resistance mediated by an I117V substitution in the neuraminidase of subtype H5N1 avian influenza a viruses. J Infect Dis 207:89–97
448. Kim DS, Son KY, Koo KM, Kim JY, Alfajaro MM, Park JG, Hosmillo M, Soliman M, Baek YB, Cho EH, Lee JH, Kang MI, Goodfellow I, Cho KO (2016) Porcine sapelovirus uses α2,3-linked Sialic acid on GD1a ganglioside as a receptor. J Virol 90(8):4067–4077

449. Han L, Tan M, Xia M, Kitova EN, Jiang X, Klassen JS (2014) Gangliosides are ligands for human noroviruses. J Am Chem Soc 136(36):12631–12637
450. Tan M, Jin M, Xie H, Duan Z, Jiang X, Fang Z (2008) Outbreak studies of a GII-3 and a GII-4 norovirus revealed an association between HBGA phenotypes and viral infection. J Med Virol 80:1296–1301
451. Wegener H, Mallagaray Á, Schöne T, Peters T, Lockhauserbäumer J, Yan H, Uetrecht C, Hansman GS, Taube S (2017) Human norovirus GII.4(MI001) P dimer binds fucosylated and sialylated carbohydrates. Glycobiology 27(11):1027–1037
452. Huang P, Farkas T, Zhong W, Tan M, Thornton S, Morrow AL, Jiang X (2005) Norovirus and histo-blood group antigens: demonstration of a wide spectrum of strain specificities and classification of two major binding groups among multiple binding patterns. J Virol 79:6714–6722
453. Song Y, Fan Z, Zuo Y, Wei H, Hu B, Chen M, Qiu R, Xue J, Wang F (2017) Binding of rabbit hemorrhagic disease virus-like particles to host histo-blood group antigens is blocked by antisera from experimentally vaccinated rabbits. Arch Virol 162(11):3425–3430.438
454. Rydell GE, Nilsson J, Rodriguez-Diaz J, Ruvoën-Clouet N, Svensson L, Le Pendu J, Larson G (2009) Human noroviruses recognize sialyl Lewis x neoglycoprotein. Glycobiology 19 (3):309–320
455. Hamel R, Dejarnac O, Wichit S, Ekchariyawat P, Neyret A, Luplertlop N, Perera-Lecoin M, Surasombatpattana P, Talignani L, Thomas F, Cao-Lormeau VM, Choumet V, Briant L, Desprès P, Amara A, Yssel H, Missé D (2015) Biology of Zika virus infection in human skin cells. J Virol 89:8880–8896
456. Cao-Lormeau VM, Blake A, Mons S, Lastère S, Roche C, Vanhomwegen J (2016) Guillain-Barre syndrome outbreak associated with Zika virus infection in French Polynesia: a case-control study. Lancet 387:1531–1539
457. Nico D, Conde L, Rivera-Correa JL, Vasconcelos-Dos-Santos A, Mesentier-Louro L, Freire-de-Lima L, Arruda MB, Freire-de-Lima CG, Ferreira ODC Jr, Lopes Moreira ME, Zin AA, Vasconcelos ZFM, Otero RM, Palatnik-de-Sousa CB, Tanuri A, Todeschini AR, Savino W, Rodriguez A, Morrot A (2018) Prevalence of IgG autoantibodies against GD3 Ganglioside in acute Zika virus infection. Front Med (Lausanne) 5:25
458. Tang H, Hammack C, Ogden SC, Wen Z, Qian X, Li Y (2016) Zika virus infects human cortical neural progenitors and attenuates their growth. Cell Stem Cell 18:587–590
459. Morris G, Barichello T, Stubbs B, Köhler CA, Carvalho AF, Maes M (2017) Zika virus as emerging neuropathogen: mechanisms of neurovirulence and neuro-immune interactions. Mol Neurobiol 55(5):4160–4184
460. Bowes T, Wagner ER, Boffey J, Nicholl D, Cochrane L, Benboubetra M (2002) Tolerance to self gangliosides is the major factor restricting the antibody response to lipopolysaccharide core oligosaccharides in *Campylobacter jejuni* strains associated with Guillain-Barre syndrome. Infect Immun 70:5008–5018
461. Melo CFOR, Delafiori J, de Oliveira DN, Guerreiro TM, Esteves CZ, Lima EO, Pando-Robles V, Catharino RR (2017) Zika-Unicamp network. Serum metabolic alterations upon Zika infection. Front Microbiol 8:1954
462. Mazzon M, Mercer J (2014) Lipid interactions during virus entry and infection. Cell Microbiol 16:1493–1502
463. Fredman P (1998) The role of antiglycolipid antibodies in neurological disorders. Ann N Y Acad Sci 845:341–352
464. Panda S, Tripathi M (2002) Anti-GQ1b IgG antibody syndrome: clinical and immunological range. J Neurol Neurosurg Psychiatry 72:418–419
465. Yuki N (2001) Infectious origins of, and molecular mimicry in, Guillain-Barre and fisher syndromes. Lancet Infect Dis 1:29–37
466. Mitsui Y, Kusunoki S, Hiruma S, Akamatsu M, Kihara M, Hashimoto S, Takahashi M (1999) Sensorimotor polyneuropathy associated with chronic lymphocytic leukemia, IgM

antigangliosides antibody and human T-cell leukemia virus I infection. Muscle Nerve 22:1461–1465
467. Jacobs BC, van Doorn PA, Schmitz PI, Tio-Gillen AP, Herbrink P, Visser LH, Hooijkass H, van der Meche FG (1996) Campylobacter jejuni infections and anti-GM1 antibodies in Guillain-Barre syndrome. Ann Neurol 40:181–187
468. Takigawa T, Yasuda H, Kikkawa R, Shigeta Y, Saida T, Kitasato H (1995) Antibodies against GM1 ganglioside affect K+ and Na+ currents in isolated rat myelinated nerve fibers. Ann Neurol 37:436–442
469. Rosenbluth J, Liang WL, Schiff R, Dou WK (1997) Spinal cord dysmyelination induced in vivo by IgM antibodies to three different myelin glycolipids. Glia 19:58–66
470. Yoshikawa T, Suzuki K, Suga S, Miyata E, Yamamoto H, Hosokawa T, Kumagai T, Asano Y (2000) Immune response to gangliosides in a case of Guillain-Barré syndrome after varicella. Arch Dis Child 83(2):172–173
471. Ang CW, Jacobs BC, Brandenburg AH, Laman JD, van der Meché FG, Osterhaus AD, van Doorn PA (2000) Cross-reactive antibodies against GM2 and CMV-infected fibroblasts in Guillain-Barré syndrome. Neurology 54(7):1453–1458
472. Wang L, Shao C, Yang C, Kang X, Zhang G (2017) Association of anti-gangliosides antibodies and anti-CMV antibodies in Guillain-Barré syndrome. Brain Behav 7(5):e00690
473. Tan M, Jiang X (2010) Norovirus gastroenteritis, carbohydrate receptors, and animal models. PLoS Pathog 6:e1000983
474. Prasad BV, Hardy ME, Dokland T, Bella J, Rossmann MG, Estes MK (1999) X-ray crystallographic structure of the Norwalk virus capsid. Science 286:287
475. Guix S, Asanaka M, Katayama K, Crawford SE, Neill FH, Atmar RL, Estes MK (2007) Norwalk virus RNA is infectious in mammalian cells. J Virol 81:12238–12248
476. Han L, Zheng R, Richards MR, Tan M, Kitova EN, Jiang X, Klassen JS (2018) Quantifying the binding stoichiometry and affinity of Histo-blood group antigen oligosaccharides for human Noroviruses. Glycobiology 28(7):488–498
477. Han L, Kitov PI, Kitova EN, Tan M, Wang L, Xia M, Jiang X, Klassen JS (2013) Affinities of recombinant norovirus P dimers for human blood group antigens. Glycobiology 23:276–285
478. Rydell GE, Dahlin AB, Höök F, Larson G (2009) QCM-D studies of human norovirus VLPs binding to glycosphingolipids in supported lipid bilayers reveal strain-specific characteristics. Glycobiology 19:1176–1184
479. Bally M, Rydell GE, Zahn R, Nasir W, Eggeling C, Breimer ME, Svensson L, Höök F, Larson G (2012) Norovirus GII.4 virus-like particles recognize galactosylceramides in domains of planar supported lipid bilayers. Angew Chem Int Ed 51:12020–12024
480. Fiege B, Rademacher C, Cartmell J, Kitov PI, Parra F, Peters T (2012) Molecular details of the recognition of blood group antigens by a human norovirus as determined by STD NMR spectroscopy. Angew Chem Int Ed 51:928–932
481. Stuart AD, Brown TDK (2007) Alpha2,6-linked sialic acid acts as a receptor for feline calicivirus. J Gen Virol 88:177–186
482. Kim DS, Hosmillo M, Alfajaro MM, Kim JY, Park JG, Son KY, Ryu EH, Sorgeloos F, Kwon HJ, Park SJ, Lee WS, Cho D, Kwon J, Choi JS, Kang MI, Goodfellow I, Cho KO (2014) Both α2,3- and α2,6-linked sialic acids on O-linked glycoproteins act as functional receptors for porcine Sapovirus. PLoS Pathog 10(6):e1004267
483. Green KY (2007) Caliciviridae: the Noroviruses. In: Knipe DM, Howley PM (eds) Fields virology, 5th edn. Lippincott Williams and Wilkins, Philadelphia, PA
484. Farkas T, Cross RW, Hargitt E 3rd, Lerche NW, Morrow AL, Sestak K (2010) Genetic diversity and histo-blood group antigen interactions of rhesus enteric caliciviruses. J Virol 84:8617–8625
485. Shirato-Horikoshi H, Ogawa S, Wakita T, Takeda N, Hansman GS (2007) Binding activity of norovirus and sapovirus to histo-blood group antigens. Arch Virol 152:457–461
486. Bank-Wolf BR, König M, Thiel HJ (2010) Zoonotic aspects of infections with noroviruses and sapoviruses. Vet Microbiol 140:204–212

487. Green KY (2007) Caliciviridae. In: Knipe DM, Howley PM (eds) Fields Virology, vol 1, 5th edn. Lippincott Williams & Wilkins, Philadelphia, PA, pp 949–980
488. Makino A, Shimojima M, Miyazawa T, Kato K, Tohya Y, Akashi H (2006) Junctional adhesion molecule 1 is a functional receptor for feline calicivirus. J Virol 80:4482–4490
489. Bomsel M, Alfsen A (2003) Entry of viruses through the epithelial barrier: pathogenic trickery. Nat Rev Mol Cell Biol 4:57–68
490. Blanco LP (2008) Gangliosides' role in cellular interactions with microbes and immune cells activation. In: Sasaki D (ed) Glycolipids: new research, vol 1. Nova Science Publishers, Inc., New York, NY, pp 29–52
491. WHO (2009) Dengue: guidelines for diagnosis, treatment, prevention, and control. Spec Program Res Train Trop Dis:147
492. Uno N, Ross TM (2018) Dengue virus and the host innate immune response. Emerg Microbes Infect 7(1):167
493. Kuhn RJ et al (2002) Structure of dengue virus: implications for flavivirus organization, maturation, and fusion. Cell 108:717–725
494. Flamand M (1999) Dengue virus type 1 nonstructural glycoprotein NS1 is secreted from mammalian cells as a soluble hexamer in a glycosylation-dependent fashion. J Virol 73:6104–6110
495. Hidari KI, Suzuki T (2011) Dengue virus receptor. Trop Med Health 39(4 Suppl):37–43.7
496. Kato D, Era S, Watanabe I, Arihara M, Sugiura N, Kimata K, Suzuki Y, Morita K, Hidari KIPJ, Suzuki T (2010) Antiviral activity of chondroitin sulphate E targeting dengue virus envelope protein. Antivir Res 88:236–243
497. Aoki C, Hidari KI, Itonori S, Yamada A, Takahashi N, Kasama T, Hasebe F, Islam MA, Hatano K, Matsuoka K, Taki T, Guo CT, Takahashi T, Sakano Y, Suzuki T, Miyamoto D, Sugita M, Terunuma D, Morita K, Suzuki Y (2006) Identification and characterization of carbohydrate molecules in mammalian cells recognized by dengue virus type 2. J Biochem 139(3):607–614
498. Wichit S, Jittmittraphap A, Hidari KI, Thaisomboonsuk B, Petmitr S, Ubol S, Aoki C, Itonori S, Morita K, Suzuki T, Suzuki Y, Jampangern W (2011) Dengue virus type 2 recognizes the carbohydrate moiety of neutral glycosphingolipids in mammalian and mosquito cells. Microbiol Immunol 55(2):135–140
499. Miller JL, de Wet BJ, Martinez-Pomares L, Radcliffe CM, Dwek RA, Rudd PM, Gordon S (2008) The mannose receptor mediates dengue virus infection of macrophages. PLoS Pathog 4:e17
500. Watson AA, Lebedev AA, Hall BA, Fenton-May AE, Vagin AA, Dejnirattisai W, Felce J, Mongkolsapaya J, Palma AS, Liu Y, Feizi T, Screaton GR, Murshudov GN, O'Callaghan CA (2011) Structural flexibility of the macrophage dengue virus receptor CLEC5A: implications for ligand binding and signaling. J Biol Chem 286(27):24208–24218
501. Tio P-H, Jong W-W, Cardosa MJ (2005) Two dimensional VOPBA reveals laminin receptor (LAMR1) interaction with dengue virus serotypes 1, 2 and 3. Virol J 2:25
502. Chen YC, Wang SY, King CC (1999) Bacterial lipopolysaccharide inhibits dengue virus infection of primary human monocytes/macrophages by blockade of virus entry via a CD14-dependent mechanism. J Virol 73:2650–2657
503. Littaua R, Kurane I, Ennis F (1990) Human IgG fc receptor II mediates antibody-dependent enhancement of dengue virus infection. J Immunol 144:3183–3186
504. Welsch S et al (2009) Composition and three-dimensional architecture of the dengue virus replication and assembly sites. Cell Host Microbe 5:365–375
505. Nasirudeen A et al (2011) RIG-I, MDA5 and TLR3 synergistically play an important role in restriction of dengue virus infection. PLoS Negl Trop Dis 5:e926
506. Sun B et al (2017) Dengue virus activates cGAS through the release of mitochondrial DNA. *Sci Rep* **7**:3594
507. Sharon N (2006) Carbohydrates as future anti-adhesion drugs for infectious diseases. Biochim Biophys Acta 1760:527–537

508. Karlsson KA (1989) Animal glycosphingolipids as membrane attachment sites for bacteria. Annu Rev Biochem 58:309–350
509. Ofek I, Doyle RJ (1994) Bacterial lectins as adhesins. In: Bacterial adhesion to cells and tissues. Chapman & Hall, New York, NY, pp 94–135
510. Karmali MA, Gannon V, Sargeant JM (2010) Verocytotoxin-producing Escherichia coli (VTEC). Vet Microbiol 140(3–4):360–370
511. Kouzel IU, Pohlentz G, Schmitz JS, Steil D, Humpf HU, Karch H, Müthing J (2017) Shiga toxin glycosphingolipid receptors in human Caco-2 and HCT-8 colon epithelial cell lines. Toxins (Basel) 9(11):E338
512. Müthing J, Meisen I, Zhang W, Bielaszewska M, Mormann M, Bauerfeind R, Schmidt MA, Friedrich AW, Karch H (2012) Promiscuous Shiga toxin 2e and its intimate relationship to Forssman. Glycobiology 22:849–862
513. Steil D, Schepers CL, Pohlentz G, Legros N, Runde J, Humpf HU, Karch H, Müthing J (2015) Shiga toxin glycosphingolipid receptors of Vero-B4 kidney epithelial cells and their membrane microdomain lipid environment. J Lipid Res 56:2322–2336
514. Bergan J, Dyve Lingelem AB, Simm R, Skotland T, Sandvig K (2015) Shiga toxins. Toxicon 60(6):1085e107
515. Raa H, Grimmer S, Schwudke D, Bergan J, Walchli S, Skotland T, Shevchenko A, Sandvig K (2009) Glycosphingolipid requirements for endosome-to-golgi transport of Shiga toxin. Traffic 10(7):868e82
516. Lindberg AA, Brown JE, Strömberg N, Westling-Ryd M, Schultz JE, Karlsson KA (1987) Identification of the carbohydrate receptor for Shiga toxin produced by Shigella dysenteriae type 1. J Biol Chem 262:1779–1785
517. Sacerdoti F, Amaral MM, Aisemberg J, Cymeryng CB, Franchi AM, Ibarra C (2015) Involvement of hypoxia and inflammation in early pregnancy loss mediated by Shiga toxin type 2. Placenta 36(6):674–680
518. Soltyk AM, MacKenzie CR, Wolski VM, Hirama T, Kitov PI, Bundle DR, Brunton JL (2002) A mutational analysis of the globotriaosylceramide-binding sites of verotoxin VT1. J Biol Chem 277(7):5351–5359
519. Ling H, Boodhoo A, Hazes B, Cummings MD, Armstrong GD, Brunton JL, Read RJ (1998) Structure of the Shiga-like toxin I B-pentamer complexed with an analogue of its receptor Gb3. Biochemistry 37(7):1777–1788
520. Tam P, Mahfoud R, Nutikka A, Khine AA, Binnington B, Paroutis P, Lingwood C (2008) Differential intracellular transport and binding of verotoxin 1 and verotoxin 2 to globotriaosylceramide-containing lipid assemblies. J Cell Physiol 216(3):750–763
521. Maloney MD, Lingwood CA (1994) CD19 has a potential CD77 (globotriaosyl ceramide)-binding site with sequence similarity to verotoxin B-subunits: implications of molecular mimicry for B cell adhesion and enterohemorrhagic *Escherichia coli* pathogenesis. J Exp Med 180(1):191–201
522. D'Angelo G, Capasso S, Sticco L, Russo D (2013) Glycosphingolipids: synthesis and functions. FEBS J 280(24):6338–6353
523. Saxena SK, O'Brien AD, Ackerman EJ (1989) Shiga toxin, Shiga-like toxin II variant, and ricin are all single-site RNA N-glycosidases of 28 S RNA when microinjected into Xenopus oocytes. J Biol Chem 264(1):596–601
524. Mori T, Kiyokawa N, Katagiri YU, Taguchi T, Suzuki T, Sekino T, Sato N, Ohmi K, Nakajima H, Takeda T, Fujimoto J (2000) Globotriaosyl ceramide (CD77/Gb3) in the glycolipid-enriched membrane domain participates in B-cell receptor-mediated apoptosis by regulating lyn kinase activity in human B cells. Exp Hematol 28(11):1260–1268
525. Moonens K, Van den Broeck I, De Kerpel M, Deboeck F, Raymaekers H, Remaut H, De Greve H (2015) Structural and functional insight into the carbohydrate receptor binding of F4 fimbriae-producing enterotoxigenic *Escherichia coli*. J Biol Chem 290(13):8409–8419
526. Bäckhed F, Alsén B, Roche N, Angström J, von Euler A, Breimer ME, Westerlund-Wikström B, Teneberg S, Richter-Dahlfors A (2002) Identification of target tissue

glycosphingolipid receptors for uropathogenic, F1C-fimbriated *Escherichia coli* and its role in mucosal inflammation. J Biol Chem 277(20):18198–18205
527. Willemsen PT, de Graaf FK (1992) Age and serotype dependent binding of K88 fimbriae to porcine intestinal receptors. Microb Pathog 12:367–375
528. Erickson AK, Willgohs JA, McFarland SY, Benfield DA, Francis DH (1992) Identification of two porcine brush border glycoproteins that bind the K88ac adhesin of *Escherichia coli* and correlation of these glycoproteins with the adhesive phenotype. Infect Immun 60:983–988
529. Erickson AK, Baker DR, Bosworth BT, Casey TA, Benfield DA, Francis DH (1994) Characterization of porcine intestinal receptors for the K88ac fimbrial adhesin of *Escherichia coli* as mucin-type sialoglycoproteins. Infect Immun 62:5404–5410
530. Grange PA, Erickson AK, Anderson TJ, Francis DH (1998) Characterization of the carbohydrate moiety of intestinal mucin-type sialoglycoprotein receptors for the K88ac fimbrial adhesin of *Escherichia coli*. Infect Immun 66:1613–1162
531. Grange PA, Mouricout MA (1996) Transferrin associated with the porcine intestinal mucosa is a receptor specific for K88ab fimbriae of *Escherichia coli*. Infect Immun 64:606–610
532. Grange PA, Erickson AK, Levery SB, Francis DH (1999) Identification of an intestinal neutral glycosphingolipid as a phenotype-specific receptor for the K88ad fimbrial adhesin of *Escherichia coli*. Infect Immun 67:165–172
533. Coddens A, Valis E, Benktander J, Ångström J, Breimer ME, Cox E, Teneberg S (2011) Erythrocyte and porcine intestinal glycosphingolipids recognized by F4 fimbriae of enterotoxigenic *Escherichia coli*. PLoS One 6:e23309
534. Gibbons RA, Jones GW, Sellwood R (1975) An attempt to identify the intestinal receptor for the K88 adhesin by means of a haemagglutination inhibition test using glycoproteins and fractions from sow colostrum. J Gen Microbiol 86:228–240
535. Grange PA, Mouricout MA, Levery SB, Francis DH, Erickson AK (2002) Evaluation of receptor binding specificity of *Escherichia coli* K88 (F4) fimbrial adhesin variants using porcine serum transferrin and glycosphingolipids as model receptors. Infect Immun 70:2336–2343
536. Sako F, Gasa S, Makita A, Hayashi A, Nozawa S (1990) Human blood group glycosphingolipids of porcine erythrocytes. Arch Biochem Biophys 278:228–237
537. Coddens A, Diswall M, Angström J, Breimer ME, Goddeeris B, Cox E, Teneberg S (2009) Recognition of blood group ABH type 1 determinants by the FedF adhesin of F18-fimbriated *Escherichia coli*. J Biol Chem 284(15):9713–9726
538. Strömberg N, Nyholm PG, Pascher I, Normark S (1991) Saccharide orientation at the cell surface affects glycolipid receptor function. Proc Natl Acad Sci U S A 88:9340–9344
539. Lillington J, Geibel S, Waksman G (2014) Biogenesis and adhesion of type 1 and P pili. Biochim Biophys Acta 1840:2783–2793
540. Lund B, Lindberg F, Marklund BI, Normark S (1987) The PapG protein is the α-Dgalactopyranosyl-(1→4)-β-D-galactopyranose-binding adhesin of uropathogenic *Escherichia coli*. Proc Natl Acad Sci U S A 84:5898–5902
541. Marklund BI, Tennent JM, Garcia E, Hamers A, Båga M, Lindberg F, Gaastra W, Normark S (1992) Horizontal gene transfer of the *Escherichia coli* pap and prs pili operons as a mechanism for the development of tissue-specific adhesive properties. Mol Microbiol 6:2225–2242
542. Ogushi K, Wada A, Niidome T, Okuda T, Llanes R, Nakayama M, Nishi Y, Kurazono H, Smith KD, Aderem A, Moss J, Hirayama T (2004) Gangliosides act as co-receptors for *Salmonella enteritidis* FliC and promote FliC induction of human beta-defensin-2 expression in Caco-2 cells. J Biol Chem 279(13):12213–12219
543. Ogushi K, Wada A, Niidome T, Mori N, Oishi K, Nagatake T, Takahashi A, Asakura H, Makino S, Hojo H, Nakahara Y, Ohsaki M, Hatakeyama T, Aoyagi H, Kurazono H, Moss J, Hirayama T (2001) *Salmonella enteritidis* FliC (flagella filament protein) induces human beta-defensin-2 mRNA production by Caco-2 cells. J Biol Chem 276:30521–30526

544. Feldman M, Bryan R, Rajan S, Scheffler L, Brunnert S, Tang H, Prince A (1998) Role of flagella in pathogenesis of *Pseudomonas aeruginosa* pulmonary infection. Infect Immun 66:43–51
545. McNamara N, Khong A, McKemy D, Caterina M, Boyer J, Julius D, Basbaum C (2001) ATP transduces signals from ASGM1, a glycolipid that functions as a bacterial receptor. Proc Natl Acad Sci U S A 98:9086–9091
546. Wain J, Hendriksen R, Mikoleit M, Keddy K, Ochiai R (2015) Typhoid fever. Lancet 385:1136–1145
547. Spano S, Ugalde JE, Galan JE (2008) Delivery of a *Salmonella Typhi* exotoxin from a host intracellular compartment. Cell Host Microbe 3:30–38
548. Song J, Gao X, Galan JE (2013) Structure and function of the *Salmonella Typhi* chimaeric a(2) B(5) typhoid toxin. Nature 499:350–354
549. Deng L, Song J, Gao X, Wang J, Yu H, Chen X, Varki N, Naito-Matsui Y, Galán J, Varki A (2014) Host adaptation of a bacterial toxin from the human pathogen *salmonella typhi*. Cell 159:1290–1299
550. Gao X, Deng L, Stack G, Yu H, Chen X, Naito-Matsui Y, Varki A, Galán JE (2017) Evolution of host adaptation in the *Salmonella typhoid* toxin. Nat Microbiol 2(12):1592–1599
551. Hodak H, Galán J (2013) *Salmonella Typhi* homolog of bacteriophage muramidases controls typhoid toxin secretion. EMBO Rep 14:95–102
552. Chang SJ, Song J, Galán JE (2016) Receptor-mediated sorting of typhoid toxin during its export from *Salmonella Typhi*-infected cells. Cell Host Microbe 20(5):682–689
553. Cho J, Chinnapen D, Aamar E, te Welscher Y, Lencer W, Massol R (2012) Insights on the trafficking and retro-translocation of glycosphingolipid-binding bacterial toxins. Front Cell Infect Microbiol 2:51–60
554. Ravindran M, Tanner L, Wenk M (2013) Sialic acid linkage in glycosphingolipids is a molecular correlate for trafficking and delivery of extracellular cargo. Traffic 14:1182–1191
555. Sandvig K, Bergan J, Kavaliauskiene S, Skotland T (2014) Lipid requirements for entry of protein toxins into cells. Prog Lipid Res 54:1–13
556. Lencer WI, Saslowsky D (2005) Raft trafficking of AB_5 subunit bacterial toxins. Biochim Biophys Acta 1746:314–321
557. Aigal S, Claudinon J, Römer W (2015) Plasma membrane reorganization: a glycolipid gateway for microbes. Biochim Biophys Acta 1853:858–871
558. MacKenzie CR, Hirama T, Lee KK, Altman E, Young NM (1997) Quantitative analysis of bacterial toxin affinity and specificity for glycolipid receptors by surface plasmon resonance. J Biol Chem 272(9):5533–5538
559. Griffiths SL, Finkelstein RA, Critchley DR (1986) Characterization of the receptor for cholera toxin and Escherichia coli heat-labile toxin in rabbit intestinal brush borders. Biochem J 238:313–322
560. Orlandi PA, Critchley DR, Fishman PH (1994) The heat-labile enterotoxin of Escherichia coli binds to polylactosaminoglycan-containing receptors in CaCo-2 human intestinal epithelial cells. Biochemistry 33:12886–12895
561. Teneberg S, Hirst TR, Ångström J, Karlsson KA (1994) Comparison of the glycolipid-binding specificities of cholera toxin and porcine Escherichia coli heat-labile enterotoxin: identification of a receptor-active non-ganglioside glycolipid for the heat-labile toxin in infant rabbit small intestine. Glycoconj J 11:533–540
562. Jobling MG, Yang Z, Kam WR, Lencer WI, Holmes RK (2012) A single native ganglioside GM1-binding site is sufficient for cholera toxin to bind to cells and complete the intoxication pathway. MBio 3(6):e00401-12
563. Wernick NL, Chinnapen DJ, Cho JA, Lencer WI (2010) Cholera toxin: an intracellular journey into the cytosol by way of the endoplasmic reticulum. Toxins (Basel) 2:310–325
564. Taylor M, Banerjee T, Ray S, Tatulian SA, Teter K (2011) Protein-disulfide isomerase displaces the cholera toxin A1 subunit from the holotoxin without unfolding the A1 subunit. J Biol Chem 286:22090–22100

565. Teter K, Jobling MG, Holmes RK (2003) A class of mutant CHO cells resistant to cholera toxin rapidly degrades the catalytic polypeptide of cholera toxin and exhibits increased endoplasmic reticulum-associated degradation. Traffic 4:232–242
566. Gill DM, Meren R (1978) ADP-ribosylation of membrane proteins catalyzed by cholera toxin: basis of the activation of adenylate cyclase. Proc Natl Acad Sci U S A 75(7):3050–3054
567. Lencer WI, Constable C, Moe S, Rufo PA, Wolf A, Jobling MG, Ruston SP, Madara JL, Holmes RK, Hirst TR (1997) Proteolytic activation of cholera toxin and *Escherichia coli* labile toxin by entry into host epithelial cells. Signal transduction by a protease-resistant toxin variant. J Biol Chem 272:15562–15568
568. Lee CH, Masso-Welch P, Hajishengallis G, Connell TD (2011) TLR2-dependent modulation of dendritic cells by LT-IIa-B5, a novel mucosal adjuvant derived from a type II heat-labile enterotoxin. J Leukoc Biol 90:911–921
569. Mitchell JS, Brown WS, Woodside DG, Vanderslice P, McIntyre BW (2009) Clustering T-cell GM1 lipid rafts increases cellular resistance to shear on fibronectin through changes in integrin affinity and cytoskeletal dynamics. Immunol Cell Biol 87:324–336
570. Bone H, Eckholdt S, Williams NA (2002) Modulation of B lymphocyte signalling by the B subunit of Escherichia coli heat-labile enterotoxin. Int Immunol 14:647–658
571. El-Kassas S, Faraj R, Martin K, Hajishengallis G, Connell TD, Nashar T (2015) Cell clustering and delay/arrest in T-cell division implicate a novel mechanism of immune modulation by *E. coli* heat-labile enterotoxin B-subunits. Cell Immunol 295(2):150–162
572. Karmarcha M, Nashar TO (2013) *E. coli* heat-labile enterotoxin B subunit as a platform for the delivery of HIV gag p24 antigen. J Clin Cell Immunol 4:1–7
573. Medzhitov R, Preston-Hurlburt P, Janeway CA Jr (1997) A human homologue of the drosophila toll protein signals activation of adaptive immunity. Nature 388:394–397
574. Aderem A, Ulevitch RJ (2000) Toll-like receptors: critical proteins linking innate and acquired immunity. Nature 406:782–787
575. Akira S (2003) Toll-like receptors. Curr Opin Immunol 15:5–11
576. O'Neill LA (2002) Signal transduction pathways activated by the IL-1 receptor/toll-like receptor superfamily. Curr Top Microbiol Immunol 270:47–61
577. Liang S, Wang M, Tapping RI, Stepensky V, Nawar HF, Triantafilou M, Triantafilou K, Connell TD, Hajishengallis G (2007) Ganglioside GD1a is an essential coreceptor for toll-like receptor 2 signaling in response to the B subunit of type IIb enterotoxin. J Biol Chem 282 (10):7532–7542
578. Hajishengallis G, Tapping RI, Martin MH, Nawar H, Lyle EA, Russell MW, Connell TD (2005) Toll-like receptor 2 mediates cellular activation by the B subunits of type II heat-labile enterotoxins. Infect Immun 73:1343–1349
579. Holmes R, Jobling MG, Connell T (1995) In: Moss J, Iglewski B, Vaughn M, Tu AT (eds) Bacterial toxins and virulence factors in disease: handbook of natural toxins. Marcel Dekker, Inc., New York, pp 225–255
580. Fukuta S, Magnani JL, Twiddy EM, Holmes RK, Ginsburg V (1988) Comparison of the carbohydrate-binding specificities of cholera toxin and *Escherichia coli* heat-labile enterotoxins LTh-I, LT-IIa, and LT-IIb. Infect Immun 56:1748–1753
581. Beutler B, Jiang Z, Georgel P, Crozat K, Croker B, Rutschmann S, Du X, Hoebe K (2006) Genetic analysis of host resistance: toll-like receptor signaling and immunity at large. Annu Rev Immunol 24:353–389
582. Takeuchi O, Sato S, Horiuchi T, Hoshino K, Takeda K, Dong Z, Modlin RL, Akira S (2002) Cutting edge: role of toll-like receptor 1 in mediating immune response to microbial lipoproteins. J Immunol 169:10–14
583. Du JW, Zhang F, Capó-Aponte JE, Tachado SD, Zhang J, Yu FS, Sack RA, Koziel H, Reinach PS (2006) AsialoGM1-mediated IL-8 release by human corneal epithelial cells requires coexpression of TLR5. Investig Ophthalmol Vis Sci 47:4810–4818
584. Anon

585. Rogers T, Snyder S (1981) High affinity binding of tetanus toxin to mammalian brain membranes. J Biol Chem 256:2402–2407
586. Adamo R, Sokol S, Soong G, Gomez MI, Prince A (2004) Pseudomonas aeruginosa flagella activate airway epithelial cells through asialoGM1 and toll-like receptor 2 as well as toll-like receptor 5. Am J Respir Cell Mol Biol 30:627–634
587. Sorice M, Garofalo T, Misasi R, Longo A, Mattei V, Sale P, Dolo V, Gradini R, Pavan A (2001) Evidence for cell surface association between CXCR4 and ganglioside GM3 after gp120 binding in SupT1 lymphoblastoid cells. FEBS Lett 506:55–60
588. Kitamura M, Iwamori M, Nagai Y (1980) Interaction between *Clostridium botulinum* neurotoxin and gangliosides. Biochim Biophys Acta 628:328–335
589. Rummel A (2017) Two feet on the membrane: uptake of clostridial neurotoxins. Curr Top Microbiol Immunol 406:1–37
590. Knapp O, Benz R, Popoff MR (2016) Pore-forming activity of clostridial binary toxins. Biochim Biophys Acta 1858:512–525
591. Nagahama M, Takehara M, Takagishi T, Seike S, Miyamoto K, Kobayashi K (2017) Cellular uptake of *Clostridium botulinum* C2 toxin requires acid sphingomyelinase activity. Infect Immun 85(4):e00966-16
592. Berntsson RP, Peng L, Dong M, Stenmark P (2013) Structure of dual receptor binding to botulinum neurotoxin B. Nat Commun 4:2058
593. Fu Z, Chen C, Barbieri JT, Kim J-JP, Baldwin MR (2009) Glycosylated SV2 and gangliosides as dual receptors for botulinum neurotoxin serotype F. Biochemistry 48:5631–5641
594. Rummel A, Häfner K, Mahrhold S, Darashchonak N, Holt M, Jahn R, Beermann S, Karnath T, Bigalke H, Binz T (2009) Botulinum neurotoxins C, E and F bind gangliosides via a conserved binding site prior to stimulation-dependent uptake with botulinum neurotoxin F utilising the three isoforms of SV2 as second receptor. J Neurochem 110(6):1942–1954
595. Rummel A, Mahrhold S, Bigalke H, Binz T (2004) The HCC-domain of botulinum neurotoxins a and B exhibits a singular ganglioside binding site displaying serotype specific carbohydrate interaction. Mol Microbiol 51:631–643
596. Fotinou C, Emsley P, Black I, Ando H, Ishida H, Kiso M, Sinha KA, Fairweather NF, Isaacs NW (2001) The crystal structure of tetanus toxin Hc fragment complexed with a synthetic GT1b analogue suggests cross-linking between ganglioside receptors and the toxin. J Biol Chem 276:32274–32281
597. Stenmark P, Dupuy J, Imamura A, Kiso M, Stevens RC (2008) 2009. Crystal structure of botulinum neurotoxin type A in complex with the cell surface co-receptor GT1b-insight into the toxin-neuron interaction. PLoS Pathog 4:e1000129
598. Millen SH, Lewallen DM, Herr AB, Iyer SS, Weiss AA (2010) Identification and characterization of the carbohydrate ligands recognized by pertussis toxin via a glycan microarray and surface plasmon resonance. Biochemistry 49(28):5954–5967
599. Benson MA, Fu Z, Kim J-JP, Baldwin MR (2011) Unique ganglioside recognition strategies for clostridial neurotoxins. J Biol Chem 286:34014–34022
600. Kozaki S, Kamata Y, Watarai S, Nishiki T, Mochida S (1998) Ganglioside GT1b as a complementary receptor component for *Clostridium botulinum* neurotoxins. Microb Pathog 25:91–99
601. Schiavo G, Matteoli M, Montecucco C (2000) Neurotoxins affecting neuroexocytosis. Physiol Rev 80:717–766
602. Schiavo G, Benfenati F, Poulai B, Rossetto O, Polverino de Laureto P, DasGupta BR, Montecucco C (1992) Tetanus and botulinum-B neurotoxins block neurotransmitter release by proteolytic cleavage of synaptobrevin. Nature (London) 359:832–835
603. Lalli G, Herreros J, Osborne SL, Montecucco C, Rossetto O, Schiavo G (1999) Functional characterisation of tetanus and botulinum neurotoxins binding domains. J Cell Sci 112:2715–2724
604. Halpern JL, Neale EA (1995) Functional characterisation of tetanus and botulinum neurotoxins binding domains. Curr Top Microbiol Immunol 195:221–241

605. Rummel A, Bad S, Alves J, Bigalke H, Binz T (2003) Two carbohydrate binding sites in the H (CC)-domain of tetanus neurotoxin are required for toxicity. J Mol Biol 326:835–847
606. Burns JR, Lambert GS, Baldwin MR (2017) Insights into the mechanisms by which Clostridial neurotoxins discriminate between Gangliosides. Biochemistry 56(20):2571–2583
607. Mutoh T, Tokuda A, Miyadai T, Hamaguchi M, Fujiki N (1995) Ganglioside GM1 binds to the Trk protein and regulates receptor function. Proc Natl Acad Sci U S A 92:5087–5091
608. Flores-Díaz M, Alape-Girón A, Clark G, Catimel B, Hirabayashi Y, Nice E, Gutiérrez JM, Titball R, Thelestam M (2005) Cellular deficiency of gangliosides causes hypersensitivity to *Clostridium perfringens* phospholipase C. J Biol Chem 280(29):26680–26689
609. Oda M, Kabura M, Takagishi T, Suzue A, Tominaga K, Urano S, Nagahama M, Kobayashi K, Furukawa K, Furukawa K, Sakurai J (2012) *Clostridium perfringens* alpha-toxin recognizes the GM1a-TrkA complex. J Biol Chem 287(39):33070–33079
610. Shimizu T, Ohtani K, Hirakawa H, Ohshima K, Yamashita A, Shiba T (2002) Complete genome sequence of *Clostridium perfringens*, an anaerobic flesh-eater. Proc Natl Acad Sci U S A 99:996–1001
611. Benson MA, Fu Z, Kim JJ, Baldwin MR (2011) Unique ganglioside recognition strategies for clostridial neurotoxins. J Biol Chem 286:34015–34022
612. Myers GS, Rasko DA, Cheung JK, Ravel J, Seshadri R, DeBoy RT (2006) Skewed genomic variability in strains of the toxigenic bacterial pathogen, *Clostridium perfringens*. Genome Res 16:1031–1040
613. Roche N, Angström J, Hurtig M, Larsson T, Borén T, Teneberg S (2004) Helicobacter pylori and complex gangliosides. Infect Immun 72(3):1519–1529
614. Angström J, Teneberg S, Milh MA, Larsson T, Leonardsson I, Olsson BM, Halvarsson MO, Danielsson D, Näslund I, Ljungh A, Wadström T, Karlsson KA (1998) The lactosylceramide binding specificity of helicobacter pylori. Glycobiology 8:297–309
615. Borén T, Falk P, Roth KA, Larson G, Normark S (1993) Attachment of helicobacter pylori to human gastric epithelium is mediated by blood group antigens. Science 262:1892–1895
616. Hirmo S, Kelm S, Schauer R, Nilsson B, Wadström T (1996) Adhesion of Helicobacter pylori strains to α-2,3-linked sialic acids. Glycoconj J 13:1005–1011
617. Johansson L, Miller-Podraza H (1998) Analysis of 3- and 6-linked sialic acids in mixtures of gangliosides using blotting to polyvinylidene difluoride membranes, binding assays and various mass spectrometry techniques with application to recognition by helicobacter pylori. Anal Biochem 265:260–268
618. Miller-Podraza H, Milh MA, Teneberg S, Karlsson KA (1997) Binding of helicobacter pylori to sialic acid-containing glycolipids of various origins separated on thin-layer chromatograms. Infect Immun 65:2480–2482
619. Miller-Podraza H, Bergström J, Milh MA, Karlsson KA (1997) Recognition of glycoconjugates by helicobacter pylori. Comparison of two sialic acid-dependent specificities based on haemagglutination and binding to human erythrocyte glycoconjugates. Glycoconj J 14:467–472
620. Miller-Podraza H, Bergström J, Teneberg S, Milh MA, Longard M, Olsson BM, Uggla L, Karlsson KA (1999) Helicobacter pylori and neutrophils. Sialic acid-dependent binding to various isolated glycoconjugates. Infect Immun 67:6309–6313
621. Roche N, Larsson T, Angström J, Teneberg S (2001) Helicobacter pylori-binding gangliosides of human gastric adenocarcinoma. Glycobiology 11:935–944
622. Saitoh T, Natomi H, Zhao WL, Okuzumi K, Sugano K, Iwamori M, Nagai Y (1991) Identification of glycolipid receptors for helicobacter pylori by TLC-immunostaining. FEBS Lett 282:385–387
623. Satin B, Del Giudice G, Della Bianca V, Dusi S, Laudanna C, Tonello F, Kelleher D, Rappuoli R, Montecucco C, Rossi F (2000) The neutrophil-activating protein (HP-NAP) of helicobacter pylori is a protective antigen and a major virulence factor. J Exp Med 191:1467–1476

624. Teneberg S (2009) The multiple carbohydrate binding specificities of helicobacter pylori. In: Lindhorst TK, Oscarson S (eds) Glycoscience and microbial adhesion/ topics in current chemistry. Springer-Verlag, Berlin Heidelberg, pp 109–138
625. Ilver D, Arnqvist A, Ögren J, Frick IM, Kersulyte D, Incecik ET, Berg DE, Covacci A, Engstrand L, Borén T (1989) Helicobacter pylori adhesin binding fucosylated histo-blood group antigens revealed by retagging. Science 279:373–377
626. Roche N, Ångström J, Hurtig M et al (2004) Helicobacter pylori and complex gangliosides. Infect Immun 72:1519–1529. https://doi.org/10.1128/IAI.72.3.1519-1529.2004
627. Siebert H-C et al (2006) a2,3/a2,6-Sialylation of N-glycans: nonsynonymous signals with marked developmental regulation in bovine reproductive tracts. Biochimie 88:399–410
628. Benktander J, Barone A, Johansson MM, Teneberg S (2018) Helicobacter pylori SabA binding gangliosides of human stomach. Virulence 9(1):738–751
629. Teneberg S, Miller-Podraza H, Lampert HC, Evans DJ Jr, Evans DG, Danielsson D, Karlsson KA (1997) Carbohydrate binding specificity of the neutrophil-activating protein of helicobacter pylori. J Biol Chem 272:19067–19071
630. Thomas RJ (2010) Receptor mimicry as novel therapeutic treatment for biothreat agents. Bioeng Bugs 1(1):17–30
631. Ariel N, Zvi A, Makarova KS, Chitlaru T, Elhanany E, Velan B et al (2003) Genome-based bioinformatic selection of chromosomal Bacillus anthracis putative vaccine candidates coupled with proteomic identification of surface-associated antigens. Infect Immun 71:4563–4579
632. Thomas RJ, Brooks T (2004) Common oligosaccharide moieties inhibit the adherence of typical and atypical respiratory pathogens. J Med Microbiol 53:833–840
633. Krivan HC, Nilsson B, Lingwood CA, Ryu H (1991) *Chlamydia trachomatis* and *Chlamydia pneumoniae* bind specifically to phosphatidylethanolamine in HeLa cells and to GalNAcβ1-4Galβ1-4Glc sequences found in asialo-GM1 and asialo-GM2. Biochem Biophys Res Commun 175:1082–1089
634. Cundell DR, Tuomanen EI (1994) Receptor specificity of adherence of *Streptococcus pneumoniae* to human type-II pneumocytes and vascular endothelial cells in vitro. Microb Pathog 17:361–374
635. Thomas RJ, Brooks T (2004) Oligosaccharide receptor mimics inhibit *Legionella pneumophila* attachment to human respiratory cell lines. Microb Pathog 36:83–92
636. Krivan HC, Ginsburg V, Roberts DD (1988) Pseudomonas aeruginosa and *Pseudomonas cepacia* isolated from cystic fibrosis patients bind specifically to gangliotetraosylceramide (Asialo GM1) and gangliotriaosylceramide (Asialo GM2). Arch Biochem Biophys 260:493–496
637. Gori AH, Ahmed K, Martinez G, Masaki H, Watanabe K, Nagatake T (1999) Mediation of attachment of *Burkholderia pseudomallei* to human pharyngeal epithelial cells by the asialoganglioside GM1-GM2 receptor complex. Am J Trop Med Hyg 61:473–475
638. Rottenberg ME, Gigliotti Rothfuchs A, Gigliotti D, Ceausu M, Une C, Levitsky V, Wigzell H (2000) Regulation and role of IFN-gamma in the innate resistance to infection with *Chlamydia pneumoniae*. J Immunol 164(9):4812–4818
639. Ahmed K, Suzuki Y, Miyamoto D, Nagatake T (2002) Asialo-GM1 and asialo-GM2 are putative adhesion molecules for *Moraxella catarrhalis*. Med Microbiol Immunol 191:5–10
640. Parente F, Cucino C, Anderloni A, Grandinetti G, Bianchi Porro G (2003) Treatment of *Helicobacter pylori* infection using novel antiadhesion compound (3′ sialyllactose sodium salt). A double blind, placebo-controlled clinical study. Helicobacter 8:252–256
641. Mock M, Fouet A (2001) Anthrax. Annu Rev Microbiol 55:647–671
642. Kern JW, Schneewind O (2008) BslA, a pXO1-encoded adhesin of *Bacillus anthracis*. Mol Microbiol 68:504–515
643. Greenfield RA, Brown BR, Hutchins JB, Iandolo JJ, Jackson R, Slater LN, Bronze MS (2002) Microbiological, biological and chemical weapons of warfare and terrorism. Am J Med Sci 323:326–340

644. Casteñeda-Roldán EI, Avelino-Flores F, Dall'Agnol M, Freer E, Cedillo L, Dornand J, , Girón JA. 2004. Adherence of Brucella to human epithelial cells and macrophages is mediated by sialic acid residues. Cell Microbiol 6, 435–445
645. Martín-Martín AI, Vizcaíno N, Fernández-Lago L (2010) Cholesterol, ganglioside GM1 and class a scavenger receptor contribute to infection by *Brucella ovis* and *Brucella canis* in murine macrophages. Microbes Infect 12(3):246–251
646. Rocha Gracia R, Casteñeda-Roldán EI, Giono-Cerezo S, Girón JA (2002) *Brucella* sp. bind to sialic acid residues on human and animal red blood cells. FEMS Microbiol Lett 231:219–224
647. Krishna K, Chitkara RK (2003) Pneumonic plague. Sem Resp Infect 18:159–167
648. Thomas RJ, Hacking A, Brooks TJG (2007) Structural modification of a base disaccharide alters antiadhesion properties towards *Yersinia pestis*. FEMS Immunol Med Microbiol 49:410–414
649. Thomas RJ, Brooks T (2006) Attachment of *Yersinia pestis* to human respiratory cell lines is inhibited by certain oligosaccharides. J Med Microbiol 55:309–315
650. Payne D, Tatham D, Williamson ED, Titball RW (1998) The pH 6 antigen of *Yersinia pestis* binds to β1-linked galactosyl residues in glycosphingolipids. Infect Immun 66:4545–4548
651. Galván EM, Chen H, Schifferli DM (2007) The Psa fimbriae of *Yersinia pestis* interact with phosphatidylcholine on alveolar epithelial cells and pulmonary surfactant. Infect Immun 75:1272–1279
652. Kienle Z, Emody L, Svanborg C, O'Toole PW (1992) Adhesive properties conferred by the plasminogen activator of *Yersinia pestis*. J Gen Microbiol 138:1679–1687
653. Kachlany SC, Planet PJ, Bhattacharjee MK, Kollia E, DeSalle R, Fine DH et al (2000) Nonspecific adherence by Actinobacillus actinomycetemcomitans requires genes widespread in bacteria and Archaea. J Bacteriol 182:6169–6176
654. Zhang P, Skurnik M, Zhang S-S, Schwartz O, Kalyanasundaram R, Bulgheresi S et al (2008) Human dendritic cell-specific intercellular adhesion molecule-grabbing nonintegrin (CD209) is a receptor for Yersinia pestis that promotes phagocytosis by dendritic cells. Infect Immun 76:2070–2079
655. Autar R, Salam Khan A, Schad M, Hacker J, Liskamp RMJ, Pieters RJ (2003) Adhesion inhibition of F1C-fimbriated Escherichia coli and Pseudomonas aeruginosa PAK and PAO by multivalent carbohydrate ligands. ChemBioChem 4:1317–1325
656. Chatterjee S, Khullar M, Shi WY (1995) Digalactosylceramide is the receptor for staphylococcal enterotoxin-B in human kidney proximal tubular cells. Glycobiology 5:327–333
657. Itakura Y, Nakamura-Tsuruta S, Kominami J, Sharon N, Kasai K, Hirabayashi J (2007) Systematic comparison of oligosaccharide specificity of Ricinus communis agglutinin I and Erythrina lectins: a search by frontal affinity chromatography. J Biochem 142:459
658. Blome MC, Schengrund CL (2008) Multivalent binding of ricin to bovine serum albumin-based neoglycoconjugates. Toxicon 51:1214–1224
659. Salomonsson E, Forsberg A, Roos N, Holz C, Maier B, Koomey M et al (2009) Functional analysis of pilin-like proteins from Francisella tularensis: complementation of type IV pilus phenotypes in Neisseria gonorrhoeae. Microbiology 155:2546–2559
660. Schulert GS, Allen LH (2006) Differential infection of mononuclear phagocytes by Francisella tularensis: role of the macrophage mannose receptor. J Leukoc Biol 80:563–571
661. Itokazu Y, Tsai YT, Yu RK (2017) Epigenetic regulation of ganglioside expression in neural stem cells and neuronal cells. Glycoconj J 34(6):749–756
662. Saito M, Sugiyama K (2002) Characterization of nuclear gangliosides in rat brain: concentration, composition, and developmental changes. Arch Biochem Biophys 398:153–159
663. Schnaar RL, Gerardy-Schahn R, Hildebrandt H (2014) Sialic acids in the brain: gangliosides and polysialic acid in nervous system development, stability, disease, and regeneration. Physiol Rev 94(2):461–518
664. Itokazu Y, Wang J, Yu RK (2018) Gangliosides in nerve cell specification. Prog Mol Biol Transl Sci 156:241–263

665. Ngamukote S, Yanagisawa M, Ariga T, Ando S, Yu RK (2007) Developmental changes of glycosphingolipids and expression of glycogenes in mouse brains. J Neurochem 103 (6):2327–2341
666. Yu RK, Tsai YT, Ariga T (2012) Functional roles of gangliosides in neurodevelopment: an overview of recent advances. Neurochem Res 37(6):1230–1244
667. McGonigal R, Barrie JA, Yao D, McLaughlin M, Cunningham ME, Rowan EG, Willison HJ (2019) Glial sulfatides and neuronal complex gangliosides are functionally interdependent in maintaining myelinating axon integrity. J Neurosci 39(1):63–77
668. Yao D, McGonigal R, Barrie JA, Cappell J, Cunningham ME, Meehan GR, Fewou SN, Edgar JM, Rowan E, Ohmi Y, Furukawa K, Furukawa K, Brophy PJ, Willison HJ (2014) Neuronal expression of GalNAc transferase is sufficient to prevent the age-related neurodegenerative phenotype of complex ganglioside-deficient mice. J Neurosci 34:880–891
669. Liour SS, Dinkins MB, Su CY, Yu RK (2005) Spatiotemporal expression of GM1 in murine medial pallial neural progenitor cells. J Comp Neurol 491(4):330–338
670. Pan B, Fromholt SE, Hess EJ, Crawford TO, Griffin JW, Sheikh KA, Schnaar RL (2005) Myelin-associated glycoprotein and complementary axonal ligands, gangliosides, mediate axon stability in the CNS and PNS: neuropathology and behavioral deficits in single- and double-null mice. Exp Neurol 195:208–217
671. Tsai YT, Yu RK (2014) Epigenetic activation of mouse ganglioside synthase genes: implications for neurogenesis. J Neurochem 128(1):101–110
672. Blaschke AJ, Staley K, Chun J (1996) Widespread programmed cell death in proliferative and postmitotic regions of the fetal cerebral cortex. Development 122:1165–1174
673. Liour SS, Kapitonov D, Yu RK (2000) Expression of gangliosides in neuronal development of P19 embryonal carcinoma stem cells. J Neurosci Res 62:363–373
674. Nakatani Y, Yanagisawa M, Suzuki Y, Yu RK (2010) Characterization of GD3 ganglioside as a novel biomarker of mouse neural stem cells. Glycobiology 20(1):78–86
675. Yanagisawa M, Yoshimura S, Yu RK (2011) Expression of GD2 and GD3 gangliosides in human embryonic neural stem cells. ASN Neuro 3(2):e00054
676. Kracun I, Rosner H, Drnovsek V, Heffer-Lauc M, Cosovic C, Lauc G (1991) Human brain gangliosides in development, aging and disease. Int J Dev Biol 35:289–295
677. Svennerholm L, Bostrom K, Fredman P, Mansson JE, Rosengren B, Rynmark BM (1989) Human brain gangliosides: developmental changes from early fetal stage to advanced age. Biochim Biophys Acta 1005:109–117
678. Norton WT, Poduslo SE (1973) Myelination in rat brain: changes in myelin composition during brain maturation. J Neurochem 21:759–773
679. Wu G, Lu ZH, Ledeen RW (1995) GM1 ganglioside in the nuclear membrane modulates nuclear calcium homeostasis during neurite outgrowth. J Neurochem 65:1419–1422
680. Li TA, Schnaar RL (2018) Congenital disorders of ganglioside biosynthesis. Prog Mol Biol Transl Sci 156:63–82
681. Harlalka GV, Lehman A, Chioza B, Baple EL, Maroofian R, Cross H et al (2013) Mutations in B4GALNT1 (GM2 synthase) underlie a new disorder of ganglioside biosynthesis. Brain 136:3618–3624
682. Magistretti PJ, Geisler FH, Schneider JS, Li PA, Fiumelli H, Sipione S (2019) Gangliosides: treatment avenues in neurodegenerative disease. Front Neurol 10:859
683. Chiricozzi E, Pome DY, Maggioni M, Di Biase E, Parravicini C, Palazzolo L et al (2017) Role of the GM1 ganglioside oligosaccharide portion in the TrkA-dependent neurite sprouting in neuroblastoma cells. J Neurochem 143:645–659
684. Hait NC, Allegood J, Maceyka M, Strub GM, Harikumar KB, Singh SK, Luo C, Marmorstein R, Kordula T, Milstien S, Spiegel S (2009) Regulation of histone acetylation in the nucleus by sphingosine-1-phosphate. Science 325:1254–1257
685. Wang J, Cheng A, Wakade C, Yu RK (2014) Ganglioside GD3 is required for neurogenesis and long-term maintenance of neural stem cells in the postnatal mouse brain. J Neurosci 34:13790–13800

686. Suzuki Y, Yanagisawa M, Yagi H, Nakatani Y, Yu RK (2010) Involvement of beta1-integrin up-regulation in basic fibroblast growth factor- and epidermal growth factor-induced proliferation of mouse neuroepithelial cells. J Biol Chem 285(24):18443–18451
687. Tsai YT, Itokazu Y, Yu RK (2016) GM1 Ganglioside is involved in epigenetic activation loci of neuronal cells. Neurochem Res 41(1-2):107–115
688. Russo D, Della Ragione F, Rizzo R, Sugiyama E, Scalabrì F, Hori K, Capasso S, Sticco L, Fioriniello S, De Gregorio R, Granata I, Guarracino MR, Maglione V, Johannes L, Bellenchi GC, Hoshino M, Setou M, D'Esposito M, Luini A, D'Angelo G (2017) Glycosphingolipid metabolic reprogramming drives neural differentiation. EMBO J 37(7):e97674
689. Gurdon JB, Bourillot PY (2001) Morphogen gradient interpretation. Nature 413 (6858):797–803
690. Gomez-Nicola D, Doncel-Perez E, Nieto-Sampedro M (2006) Regulation by GD3 of the proinflammatory response of microglia mediated by interleukin-15. J Neurosci Res 83:754–762
691. Gomez-Nicola D, Valle-Argos B, Nieto-Sampedro M (2010) Blockade of IL-15 activity inhibits microglial activation through the NFkappaB, p38, and ERK1/2 pathways, reducing cytokine and chemokine release. Glia 58(3):264–276
692. Evans MJ, Kaufman MH (1981) Establishment in culture of pluripotential cells from mouse embryos. Nature 292:154–156
693. Breimer ME, Säljö K, Barone A, Teneberg S (2017) Glycosphingolipids of human embryonic stem cells. Glycoconj J 34(6):713–723
694. Takahashi K, Yamanaka S (2006) Induction of pluripotent stem cells from mouse embryonic and adult fibroblast cultures by defined factors. Cell 126:663–676
695. Takahashi K, Tanabe K, Ohnuki M, Narita M, Ichisaka T, Tomoda K, Yamanaka S (2007) Induction of pluripotent stem cells from adult human fibroblasts by defined factors. Cell 131:861–872
696. Tang C, Lee AS, Volkmer JP, Sahoo D, Nag D, Mosley AR, Inlay MA, Ardehali R, Chavez SL, Pera RR, Behr B, Wu J, Weissman IL, Drukker M (2011) An antibody against SSEA-5 glycan on human pluripotent stem cells enables removal of teratoma-forming cells. Nat Biotechnol 29:829–834
697. Barone A, Säljö K, Benktander J, Blomqvist M, Månsson J-E, Johansson BR, Mölne J, Aspegren A, Björquist P, Breimer ME, Teneberg S (2014) Sialyl-lactotetra: a novel cell surface marker of undifferentiated human pluripotent stem cells. J Biol Chem 289:18846–18859
698. Säljö K, Barone A, Mölne J, Rydberg L, Teneberg S, Breimer ME (2017) HLA and Histo-blood group antigen expression in human pluripotent stem cells and their derivatives. Sci Rep 7(1):13072
699. Takahashi K et al (2007) Induction of pluripotent stem cells from adult human fibroblasts by defined factors. Cell 131:861–872
700. Thomson JA, Itskovitz-Eldor J, Shapiro SS, Waknitz MA, Swiergiel JJ, Marshall VS, Jones JM (1998) Embryonic stem cell lines derived from human blastocysts. Science 282:1145–1147
701. Wright AJ, Andrews PW (2009) Surface marker antigens in the characterization of human embryonic stem cells. Stem Cell Res 3:3–11
702. Lanner F, Lee KL, Sohl M et al (2010) Heparan sulfation-dependent fibroblast growth factor signaling maintains embryonic stem cells primed for differentiation in a heterogeneous state. Stem Cells 28:191–200
703. Furukawa JI, Okada K, Shinohara Y (2017) Glycomics of human embryonic stem cells and human induced pluripotent stem cells. Glycoconj J 34(6):807–815
704. Hanna JH, Saha K, Jaenisch R (2010) Pluripotency and cellular reprogramming: facts, hypotheses, unresolved issues. Cell 143:508–525
705. Grinnemo KH, Sylvén C, Hovatta O, Dellgren G, Corbascio M (2008) Immunogenicity of human embryonic stem cells. Cell Tissue Res 331:67–78

706. Solter D, Knowles BB (1978) Monoclonal antibody defining a stage-specific mouse embryonic antigen (SSEA-1). Proc Natl Acad Sci 75:5565–5569
707. Barone A et al (2013) Structural complexity of non-acid glycosphingolipids in human embryonic stem cells grown under feeder-free conditions. J Biol Chem 288:10035–10050
708. Klein HG & Anstee DJ (2014) John Wiley and Sons, Inc.
709. Clausen H, Hakomori SI (1989) ABH and related histo-blood group antigens; immunochemical differences in carrier Isotypes and their distribution. Vox Sang 56:1–20
710. Ramirez JM et al (2011) Brief report: benchmarking human pluripotent stem cell markers during differentiation into the three germ layers unveils a striking heterogeneity: all markers are not equal. Stem Cells 29:1469–1474
711. Säljö K, Barone A, Vizlin-Hodzic D, Johansson BR, Breimer ME, Funa K, Teneberg S (2016) Comparison of the glycosphingolipids of human-induced pluripotent stem cells and human embryonic stem cells. Glycobiology 27(4):291–305
712. Kannagi R, Nudelman E, Levery SB, Hakomori SI (1982) A series of human erythrocyte glycosphingolipids reacting to the monoclonal antibody directed to a developmentally regulated antigen SSEA-1. J Biol Chem 257:14865–14874
713. Robinton DA, Daley GQ (2012) The promise of induced pluripotent stem cells in research and therapy. Nature 481:295–305
714. Dahéron L, Opitz SL, Zaehres H, Lensch MW, Andrews PW, Itskovitz-Eldor J, Daley GQ (2004) LIF/STAT3 signaling fails to maintain self-renewal of human embryonic stem cells. Stem Cells 22:770–778
715. Qi X, Li TG, Hao J, Hu J, Wang J, Simmons H, Miura S, Mishina Y, Zhao GQ (2004) BMP4 supports self-renewal of embryonic stem cells by inhibiting mitogen-activated protein kinase pathways. Proc Natl Acad Sci U S A 101:6027–6032
716. Pera MF, Andrade J, Houssami S et al (2004) Regulation of human embryonic stem cell differentiation by BMP-2 and its antagonist noggin. J Cell Sci 117:1269–1280
717. Vallier L, Alexander M, Pedersen RA (2005) Activin/nodal and FGF pathways cooperate to maintain pluripotency of human embryonic stem cells. J Cell Sci 118:4495–4509
718. Sato T, Gotoh M, Kiyohara K et al (2003) Molecular cloning and characterization of a novel human beta 1,4-N-acetylgalactosaminyltransferase, beta 4GalNAc-T3, responsible for the synthesis of N,N0-diacetyllactosediamine, galNAc beta 1–4GlcNAc. J Biol Chem 278:7534–47544
719. Van den Eijnden DH, Neeleman AP, Van der Knaap WP et al (1995) Novel glycosylation routes for glycoproteins: the lacdiNAc pathway. Biochem Soc Trans 23:175–179
720. Manzella SM, Hooper LV, Baenziger JU (1996) Oligosaccharides containing beta 1,4-linked N-acetylgalactosamine, a paradigm for protein-specific glycosylation. J Biol Chem 271:12117–12120
721. Sasaki N, Shinomi M, Hirano K, Ui-Tei K, Nishihara S (2011) LacdiNAc (GalNAcβ1-4GlcNAc) contributes to self-renewal of mouse embryonic stem cells by regulating leukemia inhibitory factor/STAT3 signaling. Stem Cells 29(4):641–650
722. Sud M, Fahy E, Cotter D, Brown A, Dennis EA, Glass CK, Merrill AH Jr, Murphy RC, Raetz CR, Russell DW, Subramaniam S (2007) LMSD: LIPID MAPS structure database. Nucleic Acids Res 35:D527–D532
723. Ho MY, Yu AL, Yu J (2017) Glycosphingolipid dynamics in human embryonic stem cell and cancer: their characterization and biomedical implications. Glycoconj J 34(6):765–777
724. Barone A, Benktander J, Ångström J, Aspegren A, Björquist P, Teneberg S, Breimer ME (2013) Structural complexity of non-acid glycosphingolipids in human embryonic stem cells grown under feeder-free conditions. J Biol Chem 288(14):10035–10050
725. Liang YJ, Yang BC, Chen JM, Lin YH, Huang CL, Cheng YY, Hsu CY, Khoo KH, Shen CN, Yu J (2011) Changes in glycosphingolipid composition during differentiation of human embryonic stem cells to ectodermal or endodermal lineages. Stem Cells 29:1995–2004

726. Kannagi R, Cochran NA, Ishigami F, Hakomori S, Andrews PW, Knowles BB, Solter D (1983) Stage-specific embryonic antigens (SSEA-3 and -4) are epitopes of a unique globo-series ganglioside isolated from human teratocarcinoma cells. EMBO J 2:2355–2361
727. Kannagi R, Levery SB, Ishigami F, Hakomori S, Shevinsky LH, Knowles BB, Solter D (1983) New globoseries glycosphingolipids in human teratocarcinoma reactive with the monoclonal antibody directed to a developmentally regulated antigen, stage-specific embryonic antigen 3. J Biol Chem 258:8934–8942
728. Suila H, Pitkänen V, Hirvonen T, Heiskanen A, Anderson H, Laitinen A, Natunen S, Miller-Podraza H, Satomaa T, Natunen J, Laitinen S, Valmu L (2011) Are globoseries glycosphingolipids SSEA-3 and -4 markers for stem cells derived from human umbilical cord blood? J Mol Cell Biol 3:99–107
729. Fong CY, Peh GS, Gauthaman K, Bongso A (2009) Separation of SSEA-4 and TRA-1-60 labelled undifferentiated human embryonic stem cells from a heterogeneous cell population using magnetic-activated cell sorting (MACS) and fluorescence-activated cell sorting (FACS). Stem Cell Rev 5:72–80
730. Yu AL, Hung JT, Ho MY, Yu J (2016) Alterations of Glycosphingolipids in embryonic stem cell differentiation and development of glycan-targeting cancer immunotherapy. Stem Cells Dev 25:1532
731. Handa K, Hakomori SI (2017) Changes of glycoconjugate expression profiles during early development. Glycoconj J 34(6):693–699
732. Ben-David U, Benvenisty N (2011) The tumorigenicity of human embryonic and induced pluripotent stem cells. Nat Rev Cancer 11:268–277
733. Heimburg-Molinaro J, Lum M, Vijay G, Jain M, Almogren A, Rittenhouse-Olson K (2011) Cancer vaccines and carbohydrate epitopes. Vaccine 29:8802–8826
734. Tateno H, Toyota M, Saito S, Onuma Y, Ito Y, Hiemori K, Fukumura M, Matsushima A, Nakanishi M, Ohnuma K, Akutsu H, Umezawa A, Horimoto K, Hirabayashi J, Asashima M (2011) Glycome diagnosis of human induced pluripotent stem cells using lectin microarray. J Biol Chem 286:20345–20353
735. Manimala JC, Roach TA, Li Z, Gildersleeve JC (2007) High-throughput carbohydrate microarray profiling of 27 antibodies demonstrates widespread specificity problems. Glycobiology 17:17C–23C
736. Simons K (2016) Cell membranes: a subjective perspective. Biochim Biophys Acta 1858 (10):2569–2572
737. Jury EC, Flores-Borja F, Kabouridis PS (2007) Lipid rafts in T cell signalling and disease. Semin Cell Dev Biol 18(5):608–615
738. Zhu Y, Gumlaw N, Karman J, Zhao H, Zhang J, Jiang JL, Maniatis P, Edling A, Chuang WL, Siegel C, Shayman JA, Kaplan J, Jiang C, Cheng SH (2011) Lowering glycosphingolipid levels in CD4+ T cells attenuates T cell receptor signaling, cytokine production, and differentiation to the Th17 lineage. J Biol Chem 286(17):14787–14794
739. Nagafuku M, Okuyama K, Onimaru Y, Suzuki A, Odagiri Y, Yamashita T, Iwasaki K, Fujiwara M, Takayanagi M, Ohno I, Inokuchi J (2012) CD4 and CD8 T cells require different membrane gangliosides for activation. Proc Natl Acad Sci U S A 109(6):E336–E342
740. Balamuth F, Leitenberg D, Unternaehrer J, Mellman I, Bottomly K (2001) Distinct patterns of membrane microdomain partitioning in Th1 and Th2 cells. Immunity 15(5):729–738
741. Robinson GA, Waddington KE, Pineda-Torra I, Jury EC (2017) Transcriptional regulation of T-cell lipid metabolism: implications for plasma membrane lipid rafts and T-cell function. Front Immunol 8:1636
742. Simons K, Ikonen E (1997) Functional rafts in cell membranes. Nature 387:569–572
743. Inokuchi J, Nagafuku M, Ohno I, Suzuki A (2015) Distinct selectivity of gangliosides required for $CD4^+$ T and $CD8^+$ T cell activation. Biochim Biophys Acta 1851(1):98–106
744. Inokuchi J, Nagafuku M, Ohno I, Suzuki A (2013) Heterogeneity of gangliosides among T cell subsets. Cell Mol Life Sci 70(17):3067–3075

745. Kovacs B, Maus MV, Riley JL, Derimanov GS, Koretzky GA, June CH, Finkel TH (2002) Human $CD8^+$ T cells do not require the polarization of lipid rafts for activation and proliferation. Proc Natl Acad Sci U S A 99:15006–15011
746. Dimeloe S, Burgener AV, Grählert J, Hess C (2017) T-cell metabolism governing activation, proliferation and differentiation; a modular view. Immunology 150(1):35–44
747. McDonald G, Deepak S, Miguel L, Hall CJ, Isenberg DA, Magee AI, Butters T, Jury EC (2014) Normalizing glycosphingolipids restores function in CD4+ T cells from lupus patients. J Clin Invest 124(2):712–724
748. Nazarov-Stoica C, Surls J, Bona C, Casares S, Brumeanu TD (2009) CD28 signaling in T regulatory precursors requires p56lck and rafts integrity to stabilize the Foxp3 message. J Immunol 182(1):102–110
749. Harder T, Rentero C, Zech T, Gaus K (2007) Plasma membrane segregation during T cell activation: probing the order of domains. Curr Opin Immunol 19:470–475
750. Gómez-Móuton C, Abad JL, Mira E, Lacalle RA, Gallardo E, Jiménez-Baranda S, Illa I, Bernad A, Mañes S, Martínez AC (2001) Segregation of leading-edge and uropod components into specific lipid rafts during T cell polarization. Proc Natl Acad Sci U S A 98:9642–9647
751. Zhong L, Zhang Z, Lu X, Liu S, Chen CY, Chen ZW (2013) NSOM/QD-based visualization of GM1 serving as platforms for TCR/CD3 mediated T-cell activation. Biomed Res Int 2013:276498
752. Zhong L, Zeng G, Lu X, Wang RC, Gong G, Yan L, Huang D, Chen ZW (2009) NSOM/QD-based direct visualization of CD3-induced and CD28-enhanced nanospatial coclustering of TCR and coreceptor in nanodomains in T cell activation. PLoS One 4(6):e5945
753. Gouy H, Deterre P, Debré P, Bismuth G (1994) Cell calcium signaling via GM1 cell surface gangliosides in the human Jurkat T cell line. J Immunol 152:3271–3281
754. Wang J, Lu ZH, Gabius HJ, Rohowsky-Kochan C, Ledeen RW, Wu G (2009) Cross-linking of GM1 ganglioside by galectin-1 mediates regulatory T cell activity involving TRPC5 channel activation: possible role in suppressing experimental autoimmune encephalomyelitis. J Immunol 182:4036–4045
755. Trambley J, Bingaman AW, Lin A, Elwood ET, Waitze SY, Ha J, Durham MM, Corbascio M, Cowan SR, Pearson TC, Larsen CP (1999) Asialo GM1(+) CD8(+) T cells play a critical role in costimulation blockade-resistant allograft rejection. J Clin Invest 104:1715–1722
756. Sorice M, Parolini I, Sansolini T, Garofalo T, Dolo V, Sargiacomo M, Tai T, Peschle C, Torrisi MR, Pavan A (1997) Evidence for the existence of ganglioside-enriched plasma membrane domains in human peripheral lymphocytes. J Lipid Res 38:969–980
757. Comelli EM, Sutton-Smith M, Yan Q, Amado M, Panico M, Gilmartin T, Whisenant T, Lanigan CM, Head SR, Goldberg D, Morris HR, Dell A, Paulson JC (2006) Activation of murine CD4+ and CD8+ T lymphocytes leads to dramatic remodeling of N-linked glycans. J Immunol 177:2431–2440
758. Rossi NE, Reiné J, Pineda-Lezamit M, Pulgar M, Meza NW, Swamy M, Risueno R, Schamel WW, Bonay P, Fernández-Malavé E, Regueiro JR (2008) Differential antibody binding to the surface alphabetaTCR.CD3 complex of CD4+ and CD8+ T lymphocytes is conserved in mammals and associated with differential glycosylation. Int Immunol 20:1247–1258
759. Dong L, Hu S, Chen F, Lei X, Tu W, Yu Y, Yang L, Sun W, Yamaguchi T, Masaki Y, Umehara H (2010) Increased expression of ganglioside GM1 in peripheral CD4+ T cells correlates soluble form of CD30 in systemic lupus erythematosus patients. J Biomed Biotechnol 2010:569053
760. Shalaby KH, Martin JG (2010) Overview of asthma; the place of the T cell. Curr Opin Pharmacol 10:218–225
761. Chu JW, Sharom FJ (1990) Interleukin-2 binds to gangliosides in micelles and lipid bilayers. Biochim Biophys Acta 1028(3):205–214
762. Ravindranath MH, Gonzales A, Soh D, Nishimoto K, Tam WY, Bilchik A, Morton DL, O'Day S (2001) Interleukin-2 binds to ganglioside GD(1b). Biochem Biophys Res Commun 283 (2):369–373

763. Chu KU, Ravindranath MH, Gonzales A, Nishimoto K, Tam WT, Soh D, Bilchik A, Katopodis N, Morton DL (2000) Gangliosides as targets for immunotherapy for pancreatic adenocarcinoma. Cancer 88:1828–1836
764. Ferroni P, Lenti L, Guadagni F, Matini F, D'Agostino F, Spila A, Pontieri GM, Gazzaniga PP (1995) Possible involvement of tumour cell membrane gangliosides in platelet tumour cell interactions. Eur J Cancer 31A:79–84
765. Yuyama Y, Dohi T, Morita H, Furukawa K, Oshima M (1995) Enhanced expression of GM2/GD2 synthase mRNA in human gastrointestinal cancer. Cancer 75:1273–1280
766. Schutter EMJ, Visser JJ, van Kamp GJ, Mensdorff-Pouilly S, van Dijk W, Hilgers J, Kenemans P (1992) The utility of lipid-associated sialic acid (LASA or LSA) as a serum marker for malignancy. A review of literature. Tumor Biol 13:121–132
767. Bergelson LD (1995) Serum gangliosides as endogenous immunomodulators. Immunol Today 46:483–486
768. Zhao Z, Ciric B, Yu S, Zhang GX, Rostami A (2010) Targeting ganglioside epitope 3G11 on the surface of CD4+ T cells suppresses EAE by altering the Treg/Th17 cell balance. Int Immunol 22(10):817–826
769. Parker J, Caldini G, Krishnamurti C, Ahrens PB, Ankel H (1984) Binding of interleukin-2 to gangliosides. FEBS Lett 170:391–393
770. Kholodenko IV, Kholodenko RV, Vodovozova EL, Oleinikov VA, Polyakov NB, Molotkovskaya IM, Petrov RV (2008) Ganglioside GM1-binding sites in interleukin-4: a photoaffinity labeling study. Dokl Biochem Biophys 418:31–35
771. Lu P, Sharom F (1995) Gangliosides are potent immunosuppressors of IL-2-mediated T-cell proliferation in a low protein environment. Immunology 86(3):356–363
772. Molotkovskaya IM, Zelenova NA, Lutsenko GV, Sapozhnikov AM, Mikhalyov II, Molotkovsky JG (1999) Immunosuppressive activity of glycosphingolipids. Influence of serum factors on ganglioside inhibition of IL-4-dependent cell proliferation. Membr Cell Biol 12(6):783–791
773. Molotkovskaya IM, Kholodenko RV, Molotkovsky JG (2002) Influence of gangliosides on the IL-2- and IL-4-dependent cell proliferation. Neurochem Res 27(7–8):761–770
774. Shapiro RE, Specht CD, Collins BE, Woods AS, Cotter RJ, Schnaar RL (1997) Identification of a ganglioside recognition domain of tetanus toxin using a novel ganglioside photoaffinity ligand. J Biol Chem 272(48):30380–30386
775. Wang Y, Shen BJ, Sebald W (1997) A mixed-charge pair in human interleukin 4 dominates high-affinity interaction with the receptor alpha chain. Proc Natl Acad Sci U S A 94 (5):1657–1662
776. Cebo C, Dambrouck T, Maes E, Laden C, Strecker G, Michalski JC, Zanetta JP (2001) Recombinant human interleukins IL-1alpha, IL-1beta, IL-4, IL-6, and IL-7 show different and specific calcium-independent carbohydrate-binding properties. J Biol Chem 276 (8):5685–5691
777. Fukushima K, Hara-Kuge S, Ohkura T, Seko A, Ideo H, Inazu T, Yamashita K (1997) Lectin-like characteristics of recombinant human interleukin-1β recognizing Glycans of the glycosylphosphatidylinositol anchor. J Biol Chem 272:10579–10584
778. Zanetta JP, Vitiello F, Vincendon G (1980) Gangliosides from rat cerebellum: demonstration of considerable heterogeneity using a new solvent for thin layer chromatography. Lipids 15:1055–1061
779. Uemura S, Go S, Shishido F, Inokuchi J (2014) Expression machinery of GM4: the excess amounts of GM3/GM4S synthase (ST3GAL5) are necessary for GM4 synthesis in mammalian cells. Glycoconj J 31(2):101–108
780. Page TH, Lali FV, Foxwell BMJ (1995) Interleukin-7 activates p56lck and p59fyn, two tyrosine kinases associated with the p90 interleukin-7 receptor in primary human T cells. Eur J Immunol 25:2956–2960
781. Zanetta JP, Alonso C, Michalski JC (1996) Interleukin 2 is a lectin that associates its receptor with the T-cell receptor complex. Biochem J 318:49–53

782. Chu WK, Sharom FJ (1993) Gangliosides inhibit T-lymphocyte proliferation by preventing the interaction of interleukin-2 with its cell surface receptors. Immunology 79:10–17
783. Licinio J, Wong ML (1996) Interleukin 1 beta and fever. Nat Med 2:1314–1315
784. Tomozawa Y, Inoue T, Satoh M (1995) Expression of type I interleukin-1 receptor mRNA and its regulation in cultured astrocytes. Neurosci Lett 195:57–60
785. Ozawa H, Kotani M, Kawashima I, Numata M, Ogawa T, Terashima T, Tai T (1993) Generation of a monoclonal antibody specific for ganglioside GM4: evidence for GM4 expression on astrocytes in chicken cerebellum. J Biochem (Tokyo) 114:5–8
786. Chu JW, Sharom FJ (1995) Gangliosides interact with interleukin-4 and inhibit interleukin-4-stimulated helper T-cell proliferation. Immunology 84(3):396–403
787. Zanetta JP, Maschke S, Badache A (1993) In: Gabius H-J (ed) Lectins and Glycobiology. Springer-Verlag, Heidelberg, pp 182–187
788. Kjeldsen T, Clausen H, Hirohashi S, Ogawa T, Iijima H, Hakomori S (1988) Preparation and characterization of monoclonal antibodies directed to the tumor-associated O-linked Sialosyl-2→6 α-N-Acetylgalactosaminyl (Sialosyl-Tn) epitope. Cancer Res 48:2214–2220
789. Silva-Filho AF, Sena WLB, Lima LRA, Carvalho LVN, Pereira MC, Santos LGS, Santos RVC, Tavares LB, Pitta MGR, Rêgo MJBM (2017) Glycobiology modifications in Intratumoral hypoxia: the breathless side of Glycans interaction. Cell Physiol Biochem 41 (5):1801–1829
790. Ferreira IG, Pucci M, Venturi G, Malagolini N, Chiricolo M, Dall'Olio F (2018) Glycosylation as a main regulator of growth and death factor receptors signaling. Int J Mol Sci 19(2):E580

GPSR Compliance

The European Union's (EU) General Product Safety Regulation (GPSR) is a set of rules that requires consumer products to be safe and our obligations to ensure this.

If you have any concerns about our products, you can contact us on ProductSafety@springernature.com

In case Publisher is established outside the EU, the EU authorized representative is:

Springer Nature Customer Service Center GmbH
Europaplatz 3
69115 Heidelberg, Germany

Batch number: 10732044

Printed by Printforce, the Netherlands